This Books Belongs to:

IF YOU FOUND THE BOOK, PLEASE CONTACT:

TABLE OF CONTENTS:

Single Digit

Addition Fraction

Name: ____________ **Date:** ____________ **Score:** ____________

5

Start Time: ____________ **End Time:** ____________

S No. 1

Space for calculation

$\frac{5}{9} + \frac{9}{9} >$

S No. 2

Space for calculation

$\frac{6}{5} + \frac{7}{5} >$

S No. 3

Space for calculation

$\frac{4}{3} + \frac{4}{3} >$

S No. 4

Space for calculation

$\frac{9}{9} + \frac{2}{9} >$

S No. 5

Space for calculation

$\frac{7}{7} + \frac{4}{7} >$

S No. 6

Space for calculation

$\frac{4}{3} + \frac{9}{3} >$

S No. 7

Space for calculation

$\frac{4}{9} + \frac{3}{9} >$

S No. 8

Space for calculation

$\frac{5}{5} + \frac{4}{5} >$

S No. 9

Space for calculation

$\frac{8}{4} + \frac{8}{4} >$

S No. 10

Space for calculation

$\frac{7}{2} + \frac{2}{2} >$

Name: ______ Date: ______ Score: ______

5

Single Digit

Start Time: ______ End Time: ______

Addition Fraction

S No. 11

$$\frac{7}{3} + \frac{1}{3} >$$

Space for calculation

S No. 12

$$\frac{4}{2} + \frac{1}{2} >$$

Space for calculation

S No. 13

$$\frac{8}{2} + \frac{3}{2} >$$

Space for calculation

S No. 14

$$\frac{6}{1} + \frac{5}{1} >$$

Space for calculation

S No. 15

$$\frac{7}{7} + \frac{5}{7} >$$

Space for calculation

S No. 16

$$\frac{8}{9} + \frac{7}{9} >$$

Space for calculation

S No. 17

$$\frac{2}{4} + \frac{2}{4} >$$

Space for calculation

S No. 18

$$\frac{3}{7} + \frac{7}{7} >$$

Space for calculation

S No. 19

$$\frac{2}{5} + \frac{2}{5} >$$

Space for calculation

S No. 20

$$\frac{3}{1} + \frac{4}{1} >$$

Space for calculation

S No. 21

Space for calculation

$$\frac{86}{43} + \frac{42}{43} >$$

S No. 26

Space for calculation

$$\frac{34}{83} + \frac{63}{83} >$$

S No. 22

Space for calculation

$$\frac{58}{38} + \frac{59}{38} >$$

S No. 27

Space for calculation

$$\frac{76}{48} + \frac{45}{48} >$$

S No. 23

Space for calculation

$$\frac{67}{17} + \frac{88}{17} >$$

S No. 28

Space for calculation

$$\frac{45}{24} + \frac{89}{24} >$$

S No. 24

Space for calculation

$$\frac{83}{66} + \frac{96}{66} >$$

S No. 29

Space for calculation

$$\frac{48}{94} + \frac{50}{94} >$$

S No. 25

Space for calculation

$$\frac{84}{63} + \frac{14}{63} >$$

S No. 30

Space for calculation

$$\frac{59}{39} + \frac{64}{39} >$$

Double Digit

Addition Fraction

Name: ____________ **Date:** ________ **Score:** ________ 5

Start Time: __________ **End Time:** __________

S No. 31

Space for calculation

$\frac{41}{50} + \frac{34}{50} >$

S No. 36

Space for calculation

$\frac{47}{56} + \frac{71}{56} >$

S No. 32

Space for calculation

$\frac{98}{54} + \frac{47}{54} >$

S No. 37

Space for calculation

$\frac{97}{44} + \frac{65}{44} >$

S No. 33

Space for calculation

$\frac{65}{15} + \frac{18}{15} >$

S No. 38

Space for calculation

$\frac{30}{10} + \frac{19}{10} >$

S No. 34

Space for calculation

$\frac{57}{71} + \frac{72}{71} >$

S No. 39

Space for calculation

$\frac{92}{69} + \frac{15}{69} >$

S No. 35

Space for calculation

$\frac{94}{71} + \frac{57}{71} >$

S No. 40

Space for calculation

$\frac{66}{14} + \frac{91}{14} >$

Double Digit

Addition Fraction

Name: ____________ **Date:** ____________ **Score:** ____________

Start Time: ____________ **End Time:** ____________

S No. 41

$\frac{94}{91} + \frac{67}{91} >$

Space for calculation

S No. 42

$\frac{80}{10} + \frac{49}{10} >$

Space for calculation

S No. 43

$\frac{46}{95} + \frac{39}{95} >$

Space for calculation

S No. 44

$\frac{98}{64} + \frac{74}{64} >$

Space for calculation

S No. 45

$\frac{77}{11} + \frac{97}{11} >$

Space for calculation

S No. 46

$\frac{23}{79} + \frac{82}{79} >$

Space for calculation

S No. 47

$\frac{64}{96} + \frac{39}{96} >$

Space for calculation

S No. 48

$\frac{24}{75} + \frac{53}{75} >$

Space for calculation

S No. 49

$\frac{19}{73} + \frac{82}{73} >$

Space for calculation

S No. 50

$\frac{49}{60} + \frac{33}{60} >$

Space for calculation

Double Digit

Addition Fraction

Name: ______ **Date:** ______ **Score:** ______
5

Start Time: ______ **End Time:** ______

S No. 51

Space for calculation

$\frac{54}{97} + \frac{87}{97} >$

S No. 56

Space for calculation

$\frac{66}{62} + \frac{31}{62} >$

S No. 52

Space for calculation

$\frac{26}{54} + \frac{88}{54} >$

S No. 57

Space for calculation

$\frac{93}{30} + \frac{34}{30} >$

S No. 53

Space for calculation

$\frac{27}{76} + \frac{31}{76} >$

S No. 58

Space for calculation

$\frac{84}{62} + \frac{43}{62} >$

S No. 54

Space for calculation

$\frac{18}{35} + \frac{24}{35} >$

S No. 59

Space for calculation

$\frac{83}{99} + \frac{72}{99} >$

S No. 55

Space for calculation

$\frac{69}{64} + \frac{48}{64} >$

S No. 60

Space for calculation

$\frac{94}{91} + \frac{43}{91} >$

Double Digit

Addition Fraction

Name: ______________ **Date:** __________ **Score:** __________

5

Start Time: __________ **End Time:** __________

S No. 61

Space for calculation

$\frac{26}{45} + \frac{29}{45}$

S No. 66

Space for calculation

$\frac{62}{17} + \frac{34}{17}$

S No. 62

Space for calculation

$\frac{39}{30} + \frac{71}{30}$

S No. 67

Space for calculation

$\frac{54}{40} + \frac{87}{40}$

S No. 63

Space for calculation

$\frac{35}{56} + \frac{23}{56}$

S No. 68

Space for calculation

$\frac{83}{39} + \frac{65}{39}$

S No. 64

Space for calculation

$\frac{62}{57} + \frac{95}{57}$

S No. 69

Space for calculation

$\frac{35}{63} + \frac{58}{63}$

S No. 65

Space for calculation

$\frac{92}{10} + \frac{83}{10}$

S No. 70

Space for calculation

$\frac{76}{40} + \frac{98}{40}$

Double Digit

Addition Fraction

Name: ______ **Date:** ______ **Score:** ______

5

Start Time: ______ **End Time:** ______

S No. 71

$\frac{32}{71} + \frac{68}{71}$ >

Space for calculation

S No. 72

$\frac{44}{34} + \frac{89}{34}$ >

Space for calculation

S No. 73

$\frac{45}{87} + \frac{43}{87}$ >

Space for calculation

S No. 74

$\frac{47}{81} + \frac{54}{81}$ >

Space for calculation

S No. 75

$\frac{47}{55} + \frac{40}{55}$ >

Space for calculation

S No. 76

$\frac{12}{81} + \frac{78}{81}$ >

Space for calculation

S No. 77

$\frac{43}{94} + \frac{70}{94}$ >

Space for calculation

S No. 78

$\frac{52}{10} + \frac{57}{10}$ >

Space for calculation

S No. 79

$\frac{70}{13} + \frac{73}{13}$ >

Space for calculation

S No. 80

$\frac{99}{26} + \frac{12}{26}$ >

Space for calculation

S No. 81

Space for calculation

$$\frac{4}{1} + \frac{8}{5} >$$

S No. 82

Space for calculation

$$\frac{9}{4} + \frac{6}{1} >$$

S No. 83

Space for calculation

$$\frac{9}{8} + \frac{6}{9} >$$

S No. 84

Space for calculation

$$\frac{9}{1} + \frac{1}{4} >$$

S No. 85

Space for calculation

$$\frac{3}{5} + \frac{5}{2} >$$

S No. 86

Space for calculation

$$\frac{9}{8} + \frac{1}{7} >$$

S No. 87

Space for calculation

$$\frac{6}{3} + \frac{8}{9} >$$

S No. 88

Space for calculation

$$\frac{5}{6} + \frac{7}{4} >$$

S No. 89

Space for calculation

$$\frac{1}{1} + \frac{7}{7} >$$

S No. 90

Space for calculation

$$\frac{2}{2} + \frac{2}{1} >$$

Single Digit Different Denominator Addition Fraction

Name: ______________________ **Date:** __________ **Score:** __________ 5

Start Time: ____________ **End Time:** ____________

S No. 91

$\frac{3}{4} + \frac{2}{3} >$

Space for calculation

S No. 96

$\frac{1}{8} + \frac{3}{7} >$

Space for calculation

S No. 92

$\frac{5}{2} + \frac{6}{9} >$

Space for calculation

S No. 97

$\frac{2}{1} + \frac{1}{8} >$

Space for calculation

S No. 93

$\frac{3}{7} + \frac{4}{1} >$

Space for calculation

S No. 98

$\frac{4}{2} + \frac{6}{7} >$

Space for calculation

S No. 94

$\frac{1}{8} + \frac{7}{7} >$

Space for calculation

S No. 99

$\frac{6}{1} + \frac{7}{5} >$

Space for calculation

S No. 95

$\frac{8}{5} + \frac{4}{3} >$

Space for calculation

S No. 100

$\frac{9}{6} + \frac{4}{3} >$

Space for calculation

S No. 101

Space for calculation

$\frac{87}{83} + \frac{79}{39}$ >

S No. 102

Space for calculation

$\frac{61}{74} + \frac{76}{55}$ >

S No. 103

Space for calculation

$\frac{48}{80} + \frac{58}{60}$ >

S No. 104

Space for calculation

$\frac{89}{33} + \frac{80}{70}$ >

S No. 105

Space for calculation

$\frac{95}{68} + \frac{45}{51}$ >

S No. 106

Space for calculation

$\frac{86}{55} + \frac{73}{72}$ >

S No. 107

Space for calculation

$\frac{51}{43} + \frac{78}{28}$ >

S No. 108

Space for calculation

$\frac{92}{80} + \frac{46}{36}$ >

S No. 109

Space for calculation

$\frac{80}{30} + \frac{29}{18}$ >

S No. 110

Space for calculation

$\frac{59}{62} + \frac{16}{11}$ >

Double Digit Different Denominator Addition Fraction

Name: ______ **Date:** ______ **Score:** ______ 5

Start Time: ______ **End Time:** ______

S No. 111

$\frac{15}{44} + \frac{43}{10}$ >

Space for calculation

S No. 116

$\frac{49}{77} + \frac{47}{17}$ >

Space for calculation

S No. 112

$\frac{60}{28} + \frac{46}{23}$ >

Space for calculation

S No. 117

$\frac{87}{18} + \frac{82}{40}$ >

Space for calculation

S No. 113

$\frac{60}{66} + \frac{68}{28}$ >

Space for calculation

S No. 118

$\frac{69}{46} + \frac{49}{45}$ >

Space for calculation

S No. 114

$\frac{47}{51} + \frac{72}{37}$ >

Space for calculation

S No. 119

$\frac{93}{15} + \frac{12}{78}$ >

Space for calculation

S No. 115

$\frac{72}{84} + \frac{65}{22}$ >

Space for calculation

S No. 120

$\frac{65}{54} + \frac{61}{58}$ >

Space for calculation

Double Digit Different Denominator

Addition Fraction

Name: ____________ **Date:** ____________ **Score:** ____________ 5

Start Time: ____________ **End Time:** ____________

S No. 121

Space for calculation

$\frac{17}{78} + \frac{91}{90} >$

S No. 122

Space for calculation

$\frac{61}{60} + \frac{25}{47} >$

S No. 123

Space for calculation

$\frac{47}{24} + \frac{89}{63} >$

S No. 124

Space for calculation

$\frac{66}{95} + \frac{30}{59} >$

S No. 125

Space for calculation

$\frac{62}{55} + \frac{17}{82} >$

S No. 126

Space for calculation

$\frac{90}{88} + \frac{62}{40} >$

S No. 127

Space for calculation

$\frac{57}{41} + \frac{64}{86} >$

S No. 128

Space for calculation

$\frac{76}{52} + \frac{33}{66} >$

S No. 129

Space for calculation

$\frac{55}{99} + \frac{98}{29} >$

S No. 130

Space for calculation

$\frac{47}{77} + \frac{25}{66} >$

Double Digit Different Denominator Addition Fraction

Name: ______________________ Date: __________ Score: __________

5

Start Time: ______________ End Time: ______________

S No. 131

Space for calculation

$$\frac{51}{61} + \frac{61}{75} >$$

S No. 136

Space for calculation

$$\frac{51}{30} + \frac{84}{63} >$$

S No. 132

Space for calculation

$$\frac{27}{46} + \frac{82}{24} >$$

S No. 137

Space for calculation

$$\frac{27}{56} + \frac{23}{70} >$$

S No. 133

Space for calculation

$$\frac{29}{45} + \frac{80}{16} >$$

S No. 138

Space for calculation

$$\frac{75}{39} + \frac{49}{55} >$$

S No. 134

Space for calculation

$$\frac{58}{32} + \frac{58}{36} >$$

S No. 139

Space for calculation

$$\frac{17}{87} + \frac{50}{87} >$$

S No. 135

Space for calculation

$$\frac{57}{91} + \frac{64}{86} >$$

S No. 140

Space for calculation

$$\frac{67}{11} + \frac{21}{81} >$$

Double Digit Different Denominator

Addition Fraction

Name: ______ **Date:** ______ **Score:** ______ 5

Start Time: ______ **End Time:** ______

S No. 141

$\frac{32}{31} + \frac{24}{15}$ >

Space for calculation

S No. 142

$\frac{31}{50} + \frac{74}{33}$ >

Space for calculation

S No. 143

$\frac{29}{30} + \frac{22}{70}$ >

Space for calculation

S No. 144

$\frac{29}{10} + \frac{95}{23}$ >

Space for calculation

S No. 145

$\frac{26}{40} + \frac{70}{21}$ >

Space for calculation

S No. 146

$\frac{95}{43} + \frac{21}{37}$ >

Space for calculation

S No. 147

$\frac{16}{26} + \frac{68}{38}$ >

Space for calculation

S No. 148

$\frac{96}{14} + \frac{26}{21}$ >

Space for calculation

S No. 149

$\frac{61}{51} + \frac{23}{17}$ >

Space for calculation

S No. 150

$\frac{17}{87} + \frac{39}{79}$ >

Space for calculation

Double Digit Different Denominator

Addition Fraction

Name: ____________ **Date:** ________ **Score:** ________ 5

Start Time: ________ **End Time:** ________

S No. 151

$\frac{59}{83} + \frac{10}{94}$ >

Space for calculation

S No. 156

$\frac{51}{31} + \frac{45}{25}$ >

Space for calculation

S No. 152

$\frac{74}{71} + \frac{43}{72}$ >

Space for calculation

S No. 157

$\frac{20}{22} + \frac{84}{79}$ >

Space for calculation

S No. 153

$\frac{72}{37} + \frac{64}{78}$ >

Space for calculation

S No. 158

$\frac{98}{88} + \frac{74}{60}$ >

Space for calculation

S No. 154

$\frac{59}{54} + \frac{17}{13}$ >

Space for calculation

S No. 159

$\frac{53}{83} + \frac{95}{49}$ >

Space for calculation

S No. 155

$\frac{42}{99} + \frac{43}{97}$ >

Space for calculation

S No. 160

$\frac{22}{40} + \frac{49}{94}$ >

Space for calculation

Double Digit Different Denominator

Addition Fraction

Name: ______________ **Date:** __________ **Score:** __________

5

Start Time: __________ **End Time:** __________

S No. 161

Space for calculation

$$\frac{77}{34} + \frac{97}{18} >$$

S No. 166

Space for calculation

$$\frac{89}{82} + \frac{88}{27} >$$

S No. 162

Space for calculation

$$\frac{74}{56} + \frac{17}{85} >$$

S No. 167

Space for calculation

$$\frac{91}{17} + \frac{81}{38} >$$

S No. 163

Space for calculation

$$\frac{28}{85} + \frac{79}{15} >$$

S No. 168

Space for calculation

$$\frac{26}{57} + \frac{63}{48} >$$

S No. 164

Space for calculation

$$\frac{78}{68} + \frac{61}{84} >$$

S No. 169

Space for calculation

$$\frac{10}{93} + \frac{97}{88} >$$

S No. 165

Space for calculation

$$\frac{93}{86} + \frac{13}{10} >$$

S No. 170

Space for calculation

$$\frac{11}{48} + \frac{35}{11} >$$

Double Digit Different Denominator Addition Fraction

Name: ______________ **Date:** ________ **Score:** ________ 5

Start Time: ________ **End Time:** ________

S No. 171

Space for calculation

$\frac{20}{18} + \frac{52}{49} >$

S No. 176

Space for calculation

$\frac{51}{10} + \frac{61}{68} >$

S No. 172

Space for calculation

$\frac{79}{19} + \frac{38}{57} >$

S No. 177

Space for calculation

$\frac{34}{60} + \frac{76}{99} >$

S No. 173

Space for calculation

$\frac{71}{20} + \frac{33}{97} >$

S No. 178

Space for calculation

$\frac{77}{34} + \frac{54}{61} >$

S No. 174

Space for calculation

$\frac{36}{53} + \frac{49}{70} >$

S No. 179

Space for calculation

$\frac{76}{84} + \frac{99}{53} >$

S No. 175

Space for calculation

$\frac{27}{72} + \frac{76}{43} >$

S No. 180

Space for calculation

$\frac{77}{59} + \frac{28}{70} >$

Double Digit Different Denominator

Addition Fraction

Name: ______________________ **Date:** __________ **Score:** ________ 5

Start Time: ____________ **End Time:** ____________

S No. 181

Space for calculation

$$\frac{58}{19} + \frac{78}{94} >$$

S No. 186

Space for calculation

$$\frac{67}{94} + \frac{68}{71} >$$

S No. 182

Space for calculation

$$\frac{89}{19} + \frac{51}{49} >$$

S No. 187

Space for calculation

$$\frac{47}{63} + \frac{19}{74} >$$

S No. 183

Space for calculation

$$\frac{98}{81} + \frac{86}{97} >$$

S No. 188

Space for calculation

$$\frac{59}{16} + \frac{32}{89} >$$

S No. 184

Space for calculation

$$\frac{12}{63} + \frac{31}{81} >$$

S No. 189

Space for calculation

$$\frac{84}{29} + \frac{34}{32} >$$

S No. 185

Space for calculation

$$\frac{34}{48} + \frac{57}{11} >$$

S No. 190

Space for calculation

$$\frac{31}{89} + \frac{67}{79} >$$

Double Digit Different Denominator

Addition Fraction

Name: ______________________ **Date:** __________ **Score:** ________ 5

Start Time: ______________ **End Time:** ______________

S No. 191

$$\frac{39}{57} + \frac{79}{87} >$$

Space for calculation

S No. 196

$$\frac{82}{45} + \frac{66}{55} >$$

Space for calculation

S No. 192

$$\frac{38}{87} + \frac{70}{48} >$$

Space for calculation

S No. 197

$$\frac{66}{55} + \frac{92}{96} >$$

Space for calculation

S No. 193

$$\frac{13}{91} + \frac{99}{98} >$$

Space for calculation

S No. 198

$$\frac{50}{70} + \frac{37}{51} >$$

Space for calculation

S No. 194

$$\frac{24}{22} + \frac{36}{67} >$$

Space for calculation

S No. 199

$$\frac{57}{69} + \frac{73}{55} >$$

Space for calculation

S No. 195

$$\frac{38}{31} + \frac{36}{33} >$$

Space for calculation

S No. 200

$$\frac{13}{51} + \frac{76}{58} >$$

Space for calculation

Double Digit
Different
Denominator
Addition
Fraction

Name: ____________ **Date:** ________ **Score:** ______
5

Start Time: __________ **End Time:** __________

S No. 201

Space for calculation

$\frac{56}{30} + \frac{16}{65}$ >

S No. 202

Space for calculation

$\frac{94}{36} + \frac{78}{62}$ >

S No. 203

Space for calculation

$\frac{77}{11} + \frac{72}{11}$ >

S No. 204

Space for calculation

$\frac{37}{92} + \frac{37}{48}$ >

S No. 205

Space for calculation

$\frac{71}{81} + \frac{21}{99}$ >

S No. 206

Space for calculation

$\frac{79}{32} + \frac{62}{48}$ >

S No. 207

Space for calculation

$\frac{23}{29} + \frac{49}{15}$ >

S No. 208

Space for calculation

$\frac{95}{39} + \frac{73}{65}$ >

S No. 209

Space for calculation

$\frac{60}{53} + \frac{60}{95}$ >

S No. 210

Space for calculation

$\frac{17}{89} + \frac{56}{90}$ >

Double Digit Different Denominator

Addition Fraction

Name: ______ **Date:** ______ **Score:** ______

5

Start Time: ______ **End Time:** ______

S No. 211

Space for calculation

$\frac{30}{39} + \frac{48}{22} >$

S No. 216

Space for calculation

$\frac{85}{37} + \frac{14}{23} >$

S No. 212

Space for calculation

$\frac{63}{78} + \frac{73}{66} >$

S No. 217

Space for calculation

$\frac{43}{22} + \frac{81}{94} >$

S No. 213

Space for calculation

$\frac{59}{77} + \frac{77}{36} >$

S No. 218

Space for calculation

$\frac{51}{85} + \frac{85}{97} >$

S No. 214

Space for calculation

$\frac{83}{52} + \frac{17}{97} >$

S No. 219

Space for calculation

$\frac{92}{72} + \frac{94}{88} >$

S No. 215

Space for calculation

$\frac{28}{57} + \frac{95}{74} >$

S No. 220

Space for calculation

$\frac{64}{65} + \frac{67}{17} >$

Double Digit Different Denominator

Addition Fraction

Name: ____________ Date: ________ Score: ________ 5

Start Time: __________ End Time: __________

S No. 221

$$\frac{60}{41} + \frac{64}{66} >$$

Space for calculation

S No. 226

$$\frac{48}{40} + \frac{29}{28} >$$

Space for calculation

S No. 222

$$\frac{32}{27} + \frac{66}{34} >$$

Space for calculation

S No. 227

$$\frac{20}{78} + \frac{64}{12} >$$

Space for calculation

S No. 223

$$\frac{57}{31} + \frac{50}{60} >$$

Space for calculation

S No. 228

$$\frac{41}{30} + \frac{11}{19} >$$

Space for calculation

S No. 224

$$\frac{51}{81} + \frac{99}{45} >$$

Space for calculation

S No. 229

$$\frac{42}{37} + \frac{68}{63} >$$

Space for calculation

S No. 225

$$\frac{40}{44} + \frac{16}{59} >$$

Space for calculation

S No. 230

$$\frac{60}{77} + \frac{60}{86} >$$

Space for calculation

Double Digit Different Denominator

Addition Fraction

Name: ______ **Date:** ______ **Score:** ______ 5

Start Time: ______ **End Time:** ______

S No. 231

$\frac{57}{54} + \frac{11}{14}$ 〉

Space for calculation

S No. 236

$\frac{14}{46} + \frac{12}{96}$ 〉

Space for calculation

S No. 232

$\frac{30}{43} + \frac{68}{37}$ 〉

Space for calculation

S No. 237

$\frac{17}{75} + \frac{76}{87}$ 〉

Space for calculation

S No. 233

$\frac{59}{21} + \frac{15}{18}$ 〉

Space for calculation

S No. 238

$\frac{66}{45} + \frac{70}{53}$ 〉

Space for calculation

S No. 234

$\frac{59}{51} + \frac{98}{29}$ 〉

Space for calculation

S No. 239

$\frac{62}{99} + \frac{98}{45}$ 〉

Space for calculation

S No. 235

$\frac{91}{61} + \frac{57}{95}$ 〉

Space for calculation

S No. 240

$\frac{52}{24} + \frac{40}{97}$ 〉

Space for calculation

Double Digit Different Denominator

Addition Fraction

Name: ____________________ **Date:** __________ **Score:** __________

5

Start Time: __________ **End Time:** __________

S No. 241

$$\frac{26}{58} + \frac{83}{18} >$$

Space for calculation

S No. 242

$$\frac{72}{87} + \frac{13}{35} >$$

Space for calculation

S No. 243

$$\frac{56}{43} + \frac{13}{10} >$$

Space for calculation

S No. 244

$$\frac{32}{72} + \frac{47}{49} >$$

Space for calculation

S No. 245

$$\frac{98}{62} + \frac{52}{19} >$$

Space for calculation

S No. 246

$$\frac{20}{67} + \frac{44}{14} >$$

Space for calculation

S No. 247

$$\frac{54}{33} + \frac{81}{72} >$$

Space for calculation

S No. 248

$$\frac{40}{71} + \frac{70}{84} >$$

Space for calculation

S No. 249

$$\frac{34}{29} + \frac{94}{49} >$$

Space for calculation

S No. 250

$$\frac{93}{59} + \frac{94}{90} >$$

Space for calculation

Double Digit Different Denominator Addition Fraction

Name: ____________ **Date:** ________ **Score:** ________
5

Start Time: __________ **End Time:** __________

S No. 251

Space for calculation

$$\frac{85}{58} + \frac{64}{71} >$$

S No. 256

Space for calculation

$$\frac{24}{49} + \frac{11}{19} >$$

S No. 252

Space for calculation

$$\frac{46}{86} + \frac{77}{89} >$$

S No. 257

Space for calculation

$$\frac{17}{19} + \frac{98}{82} >$$

S No. 253

Space for calculation

$$\frac{90}{71} + \frac{56}{44} >$$

S No. 258

Space for calculation

$$\frac{82}{60} + \frac{13}{49} >$$

S No. 254

Space for calculation

$$\frac{75}{38} + \frac{93}{19} >$$

S No. 259

Space for calculation

$$\frac{94}{80} + \frac{89}{61} >$$

S No. 255

Space for calculation

$$\frac{95}{27} + \frac{35}{97} >$$

S No. 260

Space for calculation

$$\frac{29}{15} + \frac{36}{12} >$$

Double Digit Different Denominator

Addition Fraction

Name: ______________ **Date:** __________ **Score:** ________ 5

Start Time: ____________ **End Time:** ____________

S No. 261

Space for calculation

$\frac{36}{32} + \frac{43}{10}$ >

S No. 266

Space for calculation

$\frac{85}{84} + \frac{44}{95}$ >

S No. 262

Space for calculation

$\frac{78}{14} + \frac{76}{54}$ >

S No. 267

Space for calculation

$\frac{98}{57} + \frac{54}{77}$ >

S No. 263

Space for calculation

$\frac{25}{65} + \frac{31}{81}$ >

S No. 268

Space for calculation

$\frac{73}{52} + \frac{83}{89}$ >

S No. 264

Space for calculation

$\frac{81}{34} + \frac{34}{73}$ >

S No. 269

Space for calculation

$\frac{69}{34} + \frac{73}{24}$ >

S No. 265

Space for calculation

$\frac{13}{68} + \frac{72}{40}$ >

S No. 270

Space for calculation

$\frac{18}{43} + \frac{48}{65}$ >

Double Digit Different Denominator

Addition Fraction

Name: ____________ **Date:** ____________ **Score:** ____________ 5

Start Time: ____________ **End Time:** ____________

S No. 271

$\frac{49}{69} + \frac{36}{94}$ 〉

Space for calculation

S No. 272

$\frac{27}{39} + \frac{21}{45}$ 〉

Space for calculation

S No. 273

$\frac{74}{75} + \frac{87}{50}$ 〉

Space for calculation

S No. 274

$\frac{62}{51} + \frac{44}{25}$ 〉

Space for calculation

S No. 275

$\frac{26}{17} + \frac{42}{35}$ 〉

Space for calculation

S No. 276

$\frac{77}{20} + \frac{13}{67}$ 〉

Space for calculation

S No. 277

$\frac{92}{25} + \frac{94}{87}$ 〉

Space for calculation

S No. 278

$\frac{37}{50} + \frac{25}{77}$ 〉

Space for calculation

S No. 279

$\frac{65}{88} + \frac{41}{99}$ 〉

Space for calculation

S No. 280

$\frac{88}{89} + \frac{15}{87}$ 〉

Space for calculation

Double Digit Different Denominator

Addition Fraction

Name: ____________ **Date:** ________ **Score:** ________

5

Start Time: ________ **End Time:** ________

S No. 281

Space for calculation

$$\frac{49}{45} + \frac{98}{43} >$$

S No. 286

Space for calculation

$$\frac{98}{75} + \frac{77}{82} >$$

S No. 282

Space for calculation

$$\frac{76}{78} + \frac{68}{71} >$$

S No. 287

Space for calculation

$$\frac{39}{28} + \frac{33}{93} >$$

S No. 283

Space for calculation

$$\frac{93}{35} + \frac{45}{27} >$$

S No. 288

Space for calculation

$$\frac{72}{63} + \frac{46}{68} >$$

S No. 284

Space for calculation

$$\frac{87}{29} + \frac{98}{75} >$$

S No. 289

Space for calculation

$$\frac{59}{60} + \frac{86}{84} >$$

S No. 285

Space for calculation

$$\frac{75}{79} + \frac{38}{96} >$$

S No. 290

Space for calculation

$$\frac{50}{86} + \frac{87}{90} >$$

Double Digit Different Denominator Addition Fraction

Name: ____________ **Date:** ____________ **Score:** ____________

5

Start Time: ____________ **End Time:** ____________

S No. 291

Space for calculation

$$\frac{96}{90} + \frac{24}{22} >$$

S No. 296

Space for calculation

$$\frac{76}{52} + \frac{28}{94} >$$

S No. 292

Space for calculation

$$\frac{77}{10} + \frac{43}{35} >$$

S No. 297

Space for calculation

$$\frac{55}{69} + \frac{59}{27} >$$

S No. 293

Space for calculation

$$\frac{59}{49} + \frac{70}{56} >$$

S No. 298

Space for calculation

$$\frac{69}{12} + \frac{15}{93} >$$

S No. 294

Space for calculation

$$\frac{37}{22} + \frac{49}{32} >$$

S No. 299

Space for calculation

$$\frac{67}{10} + \frac{82}{36} >$$

S No. 295

Space for calculation

$$\frac{69}{97} + \frac{37}{35} >$$

S No. 300

Space for calculation

$$\frac{16}{18} + \frac{86}{40} >$$

Double Digit Different Denominator

Addition Fraction

Name: ____________ **Date:** ________ **Score:** ________ 5

Start Time: ________ **End Time:** ________

S No. 301

$\frac{57}{41} + \frac{19}{24} >$

Space for calculation

S No. 302

$\frac{12}{71} + \frac{47}{12} >$

Space for calculation

S No. 303

$\frac{58}{17} + \frac{10}{26} >$

Space for calculation

S No. 304

$\frac{93}{76} + \frac{95}{11} >$

Space for calculation

S No. 305

$\frac{49}{83} + \frac{70}{82} >$

Space for calculation

S No. 306

$\frac{90}{81} + \frac{78}{64} >$

Space for calculation

S No. 307

$\frac{88}{28} + \frac{14}{11} >$

Space for calculation

S No. 308

$\frac{80}{43} + \frac{91}{81} >$

Space for calculation

S No. 309

$\frac{45}{25} + \frac{77}{68} >$

Space for calculation

S No. 310

$\frac{77}{43} + \frac{92}{89} >$

Space for calculation

Double Digit Different Denominator

Addition Fraction

Name: ______ **Date:** ______ **Score:** ______ 5

Start Time: ______ **End Time:** ______

S No. 311

Space for calculation

$\frac{24}{42} + \frac{97}{86} >$

S No. 316

Space for calculation

$\frac{14}{49} + \frac{27}{33} >$

S No. 312

Space for calculation

$\frac{49}{31} + \frac{97}{29} >$

S No. 317

Space for calculation

$\frac{96}{34} + \frac{99}{46} >$

S No. 313

Space for calculation

$\frac{37}{60} + \frac{61}{52} >$

S No. 318

Space for calculation

$\frac{92}{13} + \frac{45}{55} >$

S No. 314

Space for calculation

$\frac{97}{90} + \frac{93}{16} >$

S No. 319

Space for calculation

$\frac{57}{36} + \frac{33}{32} >$

S No. 315

Space for calculation

$\frac{75}{86} + \frac{51}{48} >$

S No. 320

Space for calculation

$\frac{87}{82} + \frac{26}{53} >$

Double Digit Different Denominator
Addition Fraction

Name: ____________ **Date:** ________ **Score:** ________
5

Start Time: ________ **End Time:** ________

S No. 321

$\frac{73}{79} + \frac{87}{52} >$

Space for calculation

S No. 326

$\frac{11}{87} + \frac{31}{56} >$

Space for calculation

S No. 322

$\frac{29}{82} + \frac{65}{84} >$

Space for calculation

S No. 327

$\frac{37}{92} + \frac{87}{12} >$

Space for calculation

S No. 323

$\frac{46}{46} + \frac{26}{85} >$

Space for calculation

S No. 328

$\frac{13}{57} + \frac{59}{35} >$

Space for calculation

S No. 324

$\frac{43}{25} + \frac{72}{49} >$

Space for calculation

S No. 329

$\frac{21}{11} + \frac{18}{80} >$

Space for calculation

S No. 325

$\frac{38}{91} + \frac{32}{19} >$

Space for calculation

S No. 330

$\frac{53}{97} + \frac{50}{92} >$

Space for calculation

Double Digit Different Denominator Addition Fraction

Name: ____________________ **Date:** __________ **Score:** __________
5

Start Time: ______________ **End Time:** ______________

S No. 331

Space for calculation

$$\frac{63}{95} + \frac{73}{34} \rangle$$

S No. 336

Space for calculation

$$\frac{62}{20} + \frac{83}{91} \rangle$$

S No. 332

Space for calculation

$$\frac{27}{24} + \frac{56}{94} \rangle$$

S No. 337

Space for calculation

$$\frac{25}{28} + \frac{43}{97} \rangle$$

S No. 333

Space for calculation

$$\frac{17}{23} + \frac{69}{83} \rangle$$

S No. 338

Space for calculation

$$\frac{34}{98} + \frac{86}{67} \rangle$$

S No. 334

Space for calculation

$$\frac{86}{52} + \frac{73}{93} \rangle$$

S No. 339

Space for calculation

$$\frac{26}{36} + \frac{72}{18} \rangle$$

S No. 335

Space for calculation

$$\frac{35}{22} + \frac{71}{33} \rangle$$

S No. 340

Space for calculation

$$\frac{39}{32} + \frac{34}{94} \rangle$$

Double Digit Different Denominator

Addition Fraction

Name: ______ **Date:** ______ **Score:** ______

5

Start Time: ______ **End Time:** ______

S No. 341

$\frac{92}{82} + \frac{97}{97}$ >

Space for calculation

S No. 342

$\frac{43}{97} + \frac{45}{98}$ >

Space for calculation

S No. 343

$\frac{68}{21} + \frac{26}{26}$ >

Space for calculation

S No. 344

$\frac{81}{88} + \frac{38}{69}$ >

Space for calculation

S No. 345

$\frac{13}{26} + \frac{80}{14}$ >

Space for calculation

S No. 346

$\frac{45}{90} + \frac{34}{25}$ >

Space for calculation

S No. 347

$\frac{85}{31} + \frac{13}{78}$ >

Space for calculation

S No. 348

$\frac{58}{85} + \frac{60}{10}$ >

Space for calculation

S No. 349

$\frac{93}{13} + \frac{25}{12}$ >

Space for calculation

S No. 350

$\frac{55}{15} + \frac{17}{25}$ >

Space for calculation

Single Digit

Subtraction Fraction

Name: ______________ **Date:** ________ **Score:** ______ 5

Start Time: ____________ **End Time:** ____________

S No. 351

Space for calculation

$\frac{5}{9} - \frac{9}{9} >$

S No. 352

Space for calculation

$\frac{6}{5} - \frac{7}{5} >$

S No. 353

Space for calculation

$\frac{4}{3} - \frac{4}{3} >$

S No. 354

Space for calculation

$\frac{9}{9} - \frac{2}{9} >$

S No. 355

Space for calculation

$\frac{7}{7} - \frac{4}{7} >$

S No. 356

Space for calculation

$\frac{4}{3} - \frac{9}{3} >$

S No. 357

Space for calculation

$\frac{4}{9} - \frac{3}{9} >$

S No. 358

Space for calculation

$\frac{5}{5} - \frac{4}{5} >$

S No. 359

Space for calculation

$\frac{8}{4} - \frac{8}{4} >$

S No. 360

Space for calculation

$\frac{7}{2} - \frac{2}{2} >$

Single Digit

Subtraction Fraction

Name: ____________ **Date:** ____________ **Score:** ____________ 5

Start Time: ____________ **End Time:** ____________

S No. 361

Space for calculation

$\frac{7}{3} - \frac{1}{3}$ >

S No. 362

Space for calculation

$\frac{4}{2} - \frac{1}{2}$ >

S No. 363

Space for calculation

$\frac{8}{2} - \frac{3}{2}$ >

S No. 364

Space for calculation

$\frac{6}{1} - \frac{5}{1}$ >

S No. 365

Space for calculation

$\frac{7}{7} - \frac{5}{7}$ >

S No. 366

Space for calculation

$\frac{8}{9} - \frac{7}{9}$ >

S No. 367

Space for calculation

$\frac{2}{4} - \frac{2}{4}$ >

S No. 368

Space for calculation

$\frac{3}{7} - \frac{7}{7}$ >

S No. 369

Space for calculation

$\frac{2}{5} - \frac{2}{5}$ >

S No. 370

Space for calculation

$\frac{3}{1} - \frac{4}{1}$ >

Double Digit

Subtraction Fraction

Name: ______________________ **Date:** __________ **Score:** __________

5

Start Time: ______________ **End Time:** ______________

S No. 371

$$\frac{86}{43} - \frac{42}{43} >$$

Space for calculation

S No. 376

$$\frac{34}{83} - \frac{63}{83} >$$

Space for calculation

S No. 372

$$\frac{58}{38} - \frac{59}{38} >$$

Space for calculation

S No. 377

$$\frac{76}{48} - \frac{45}{48} >$$

Space for calculation

S No. 373

$$\frac{67}{17} - \frac{88}{17} >$$

Space for calculation

S No. 378

$$\frac{45}{24} - \frac{89}{24} >$$

Space for calculation

S No. 374

$$\frac{83}{66} - \frac{96}{66} >$$

Space for calculation

S No. 379

$$\frac{48}{94} - \frac{50}{94} >$$

Space for calculation

S No. 375

$$\frac{84}{63} - \frac{14}{63} >$$

Space for calculation

S No. 380

$$\frac{59}{39} - \frac{64}{39} >$$

Space for calculation

Double Digit

Subtraction Fraction

Name: ______________________ **Date:** ____________ **Score:** ________ / 5

Start Time: ______________ **End Time:** ______________

S No. 381

$\frac{41}{50} - \frac{34}{50}$ >

Space for calculation

S No. 386

$\frac{47}{56} - \frac{71}{56}$ >

Space for calculation

S No. 382

$\frac{98}{54} - \frac{47}{54}$ >

Space for calculation

S No. 387

$\frac{97}{44} - \frac{65}{44}$ >

Space for calculation

S No. 383

$\frac{65}{15} - \frac{18}{15}$ >

Space for calculation

S No. 388

$\frac{30}{10} - \frac{19}{10}$ >

Space for calculation

S No. 384

$\frac{57}{71} - \frac{72}{71}$ >

Space for calculation

S No. 389

$\frac{92}{69} - \frac{15}{69}$ >

Space for calculation

S No. 385

$\frac{94}{71} - \frac{57}{71}$ >

Space for calculation

S No. 390

$\frac{66}{14} - \frac{91}{14}$ >

Space for calculation

Double Digit

Subtraction Fraction

Name: ____________ **Date:** ____________ **Score:** ____________ 5

Start Time: ____________ **End Time:** ____________

S No. 391

$\frac{94}{91} - \frac{67}{91} >$

Space for calculation

S No. 392

$\frac{80}{10} - \frac{49}{10} >$

Space for calculation

S No. 393

$\frac{46}{95} - \frac{39}{95} >$

Space for calculation

S No. 394

$\frac{98}{64} - \frac{74}{64} >$

Space for calculation

S No. 395

$\frac{77}{11} - \frac{97}{11} >$

Space for calculation

S No. 396

$\frac{23}{79} - \frac{82}{79} >$

Space for calculation

S No. 397

$\frac{64}{96} - \frac{39}{96} >$

Space for calculation

S No. 398

$\frac{24}{75} - \frac{53}{75} >$

Space for calculation

S No. 399

$\frac{19}{73} - \frac{82}{73} >$

Space for calculation

S No. 400

$\frac{49}{60} - \frac{33}{60} >$

Space for calculation

Double Digit

Subtraction Fraction

Name: ______________ **Date:** ________ **Score:** ______
5

Start Time: ________ **End Time:** ________

S No. 401

$\frac{54}{97} - \frac{87}{97} >$

Space for calculation

S No. 402

$\frac{26}{54} - \frac{88}{54} >$

Space for calculation

S No. 403

$\frac{27}{76} - \frac{31}{76} >$

Space for calculation

S No. 404

$\frac{18}{35} - \frac{24}{35} >$

Space for calculation

S No. 405

$\frac{69}{64} - \frac{48}{64} >$

Space for calculation

S No. 406

$\frac{66}{62} - \frac{31}{62} >$

Space for calculation

S No. 407

$\frac{93}{30} - \frac{34}{30} >$

Space for calculation

S No. 408

$\frac{84}{62} - \frac{43}{62} >$

Space for calculation

S No. 409

$\frac{83}{99} - \frac{72}{99} >$

Space for calculation

S No. 410

$\frac{94}{91} - \frac{43}{91} >$

Space for calculation

Double Digit

Subtraction Fraction

Name: ____________ **Date:** ________ **Score:** ________

5

Start Time: ________ **End Time:** ________

S No. 411

$\frac{26}{45} - \frac{29}{45} >$

Space for calculation

S No. 416

$\frac{62}{17} - \frac{34}{17} >$

Space for calculation

S No. 412

$\frac{39}{30} - \frac{71}{30} >$

Space for calculation

S No. 417

$\frac{54}{40} - \frac{87}{40} >$

Space for calculation

S No. 413

$\frac{35}{56} - \frac{23}{56} >$

Space for calculation

S No. 418

$\frac{83}{39} - \frac{65}{39} >$

Space for calculation

S No. 414

$\frac{62}{57} - \frac{95}{57} >$

Space for calculation

S No. 419

$\frac{35}{63} - \frac{58}{63} >$

Space for calculation

S No. 415

$\frac{92}{10} - \frac{83}{10} >$

Space for calculation

S No. 420

$\frac{76}{40} - \frac{98}{40} >$

Space for calculation

Double Digit

Subtraction Fraction

Name: ______ Date: ______ Score: ______

5

Start Time: ______ End Time: ______

S No. 421

$\frac{32}{71} - \frac{68}{71} >$

Space for calculation

S No. 426

$\frac{12}{81} - \frac{78}{81} >$

Space for calculation

S No. 422

$\frac{44}{34} - \frac{89}{34} >$

Space for calculation

S No. 427

$\frac{43}{94} - \frac{70}{94} >$

Space for calculation

S No. 423

$\frac{45}{87} - \frac{43}{87} >$

Space for calculation

S No. 428

$\frac{52}{10} - \frac{57}{10} >$

Space for calculation

S No. 424

$\frac{47}{81} - \frac{54}{81} >$

Space for calculation

S No. 429

$\frac{70}{13} - \frac{73}{13} >$

Space for calculation

S No. 425

$\frac{47}{55} - \frac{40}{55} >$

Space for calculation

S No. 430

$\frac{99}{26} - \frac{12}{26} >$

Space for calculation

Single Digit Different Denominator

Subtraction Fraction

Name: ______ **Date:** ______ **Score:** ______ 5

Start Time: ______ **End Time:** ______

S No. 431

$\frac{4}{1} - \frac{8}{5}$ >

Space for calculation

S No. 436

$\frac{9}{8} - \frac{1}{7}$ >

Space for calculation

S No. 432

$\frac{9}{4} - \frac{6}{1}$ >

Space for calculation

S No. 437

$\frac{6}{3} - \frac{8}{9}$ >

Space for calculation

S No. 433

$\frac{9}{8} - \frac{6}{9}$ >

Space for calculation

S No. 438

$\frac{5}{6} - \frac{7}{4}$ >

Space for calculation

S No. 434

$\frac{9}{1} - \frac{1}{4}$ >

Space for calculation

S No. 439

$\frac{1}{1} - \frac{7}{7}$ >

Space for calculation

S No. 435

$\frac{3}{5} - \frac{5}{2}$ >

Space for calculation

S No. 440

$\frac{2}{2} - \frac{2}{1}$ >

Space for calculation

Single Digit Different Denominator

Subtraction Fraction

Name: ______________________ **Date:** ____________ **Score:** __________ 5

Start Time: ______________ **End Time:** ______________

S No. 441

$\frac{3}{4} - \frac{2}{3}$ >

Space for calculation

S No. 446

$\frac{1}{8} - \frac{3}{7}$ >

Space for calculation

S No. 442

$\frac{5}{2} - \frac{6}{9}$ >

Space for calculation

S No. 447

$\frac{2}{1} - \frac{1}{8}$ >

Space for calculation

S No. 443

$\frac{3}{7} - \frac{4}{1}$ >

Space for calculation

S No. 448

$\frac{4}{2} - \frac{6}{7}$ >

Space for calculation

S No. 444

$\frac{1}{8} - \frac{7}{7}$ >

Space for calculation

S No. 449

$\frac{6}{1} - \frac{7}{5}$ >

Space for calculation

S No. 445

$\frac{8}{5} - \frac{4}{3}$ >

Space for calculation

S No. 450

$\frac{9}{6} - \frac{4}{3}$ >

Space for calculation

Double Digit Different Denominator

Subtraction Fraction

Name: ____________ **Date:** ________ **Score:** ________

5

Start Time: __________ **End Time:** __________

S No. 451

$\frac{87}{83} - \frac{79}{39} >$

Space for calculation

S No. 452

$\frac{61}{74} - \frac{76}{55} >$

Space for calculation

S No. 453

$\frac{48}{80} - \frac{58}{60} >$

Space for calculation

S No. 454

$\frac{89}{33} - \frac{80}{70} >$

Space for calculation

S No. 455

$\frac{95}{68} - \frac{45}{51} >$

Space for calculation

S No. 456

$\frac{86}{55} - \frac{73}{72} >$

Space for calculation

S No. 457

$\frac{51}{43} - \frac{78}{28} >$

Space for calculation

S No. 458

$\frac{92}{80} - \frac{46}{36} >$

Space for calculation

S No. 459

$\frac{80}{30} - \frac{29}{18} >$

Space for calculation

S No. 460

$\frac{59}{62} - \frac{16}{11} >$

Space for calculation

Double Digit Different Denominator

Subtraction Fraction

Name: ______________________ **Date:** ____________ **Score:** __________
5

Start Time: ______________ **End Time:** ______________

S No. 461

Space for calculation

$\frac{15}{44} - \frac{43}{10}$ >

S No. 466

Space for calculation

$\frac{49}{77} - \frac{47}{17}$ >

S No. 462

Space for calculation

$\frac{60}{28} - \frac{46}{23}$ >

S No. 467

Space for calculation

$\frac{87}{18} - \frac{82}{40}$ >

S No. 463

Space for calculation

$\frac{60}{66} - \frac{68}{28}$ >

S No. 468

Space for calculation

$\frac{69}{46} - \frac{49}{45}$ >

S No. 464

Space for calculation

$\frac{47}{51} - \frac{72}{37}$ >

S No. 469

Space for calculation

$\frac{93}{15} - \frac{12}{78}$ >

S No. 465

Space for calculation

$\frac{72}{84} - \frac{65}{22}$ >

S No. 470

Space for calculation

$\frac{65}{54} - \frac{61}{58}$ >

Double Digit Different Denominator

Subtraction Fraction

Name: ________ **Date:** ________ **Score:** ________ 5

Start Time: ________ **End Time:** ________

S No. 471

$\frac{17}{78} - \frac{91}{90} >$

Space for calculation

S No. 476

$\frac{90}{88} - \frac{62}{40} >$

Space for calculation

S No. 472

$\frac{61}{60} - \frac{25}{47} >$

Space for calculation

S No. 477

$\frac{57}{41} - \frac{64}{86} >$

Space for calculation

S No. 473

$\frac{47}{24} - \frac{89}{63} >$

Space for calculation

S No. 478

$\frac{76}{52} - \frac{33}{66} >$

Space for calculation

S No. 474

$\frac{66}{95} - \frac{30}{59} >$

Space for calculation

S No. 479

$\frac{55}{99} - \frac{98}{29} >$

Space for calculation

S No. 475

$\frac{62}{55} - \frac{17}{82} >$

Space for calculation

S No. 480

$\frac{47}{77} - \frac{25}{66} >$

Space for calculation

Double Digit Different Denominator

Subtraction Fraction

Name: ______________________ **Date:** ____________ **Score:** __________ 5

Start Time: ______________ **End Time:** ______________

S No. 481

$\frac{51}{61} - \frac{61}{75}$ >

Space for calculation

S No. 482

$\frac{27}{46} - \frac{82}{24}$ >

Space for calculation

S No. 483

$\frac{29}{45} - \frac{80}{16}$ >

Space for calculation

S No. 484

$\frac{58}{32} - \frac{58}{36}$ >

Space for calculation

S No. 485

$\frac{57}{91} - \frac{64}{86}$ >

Space for calculation

S No. 486

$\frac{51}{30} - \frac{84}{63}$ >

Space for calculation

S No. 487

$\frac{27}{56} - \frac{23}{70}$ >

Space for calculation

S No. 488

$\frac{75}{39} - \frac{49}{55}$ >

Space for calculation

S No. 489

$\frac{17}{87} - \frac{50}{87}$ >

Space for calculation

S No. 490

$\frac{67}{11} - \frac{21}{81}$ >

Space for calculation

Double Digit Different Denominator

Subtraction Fraction

Name: ______________________ **Date:** __________ **Score:** __________

5

Start Time: ______________ **End Time:** ______________

S No. 491

$\frac{32}{31} - \frac{24}{15}$ >

Space for calculation

S No. 496

$\frac{95}{43} - \frac{21}{37}$ >

Space for calculation

S No. 492

$\frac{31}{50} - \frac{74}{33}$ >

Space for calculation

S No. 497

$\frac{16}{26} - \frac{68}{38}$ >

Space for calculation

S No. 493

$\frac{29}{30} - \frac{22}{70}$ >

Space for calculation

S No. 498

$\frac{96}{14} - \frac{26}{21}$ >

Space for calculation

S No. 494

$\frac{29}{10} - \frac{95}{23}$ >

Space for calculation

S No. 499

$\frac{61}{51} - \frac{23}{17}$ >

Space for calculation

S No. 495

$\frac{26}{40} - \frac{70}{21}$ >

Space for calculation

S No. 500

$\frac{17}{87} - \frac{39}{79}$ >

Space for calculation

Double Digit Different Denominator

Subtraction Fraction

Name: ____________ **Date:** ____________ **Score:** ____________ 5

Start Time: ____________ **End Time:** ____________

S No. 501

$\frac{59}{83} - \frac{10}{94} >$

Space for calculation

S No. 506

$\frac{51}{31} - \frac{45}{25} >$

Space for calculation

S No. 502

$\frac{74}{71} - \frac{43}{72} >$

Space for calculation

S No. 507

$\frac{20}{22} - \frac{84}{79} >$

Space for calculation

S No. 503

$\frac{72}{37} - \frac{64}{78} >$

Space for calculation

S No. 508

$\frac{98}{88} - \frac{74}{60} >$

Space for calculation

S No. 504

$\frac{59}{54} - \frac{17}{13} >$

Space for calculation

S No. 509

$\frac{53}{83} - \frac{95}{49} >$

Space for calculation

S No. 505

$\frac{42}{99} - \frac{43}{97} >$

Space for calculation

S No. 510

$\frac{22}{40} - \frac{49}{94} >$

Space for calculation

Double Digit Different Denominator

Subtraction Fraction

Name: ____________ **Date:** ________ **Score:** ________

5

Start Time: ________ **End Time:** ________

S No. 511

$\frac{77}{34} - \frac{97}{18}$ >

Space for calculation

S No. 516

$\frac{89}{82} - \frac{88}{27}$ >

Space for calculation

S No. 512

$\frac{74}{56} - \frac{17}{85}$ >

Space for calculation

S No. 517

$\frac{91}{17} - \frac{81}{38}$ >

Space for calculation

S No. 513

$\frac{28}{85} - \frac{79}{15}$ >

Space for calculation

S No. 518

$\frac{26}{57} - \frac{63}{48}$ >

Space for calculation

S No. 514

$\frac{78}{68} - \frac{61}{84}$ >

Space for calculation

S No. 519

$\frac{10}{93} - \frac{97}{88}$ >

Space for calculation

S No. 515

$\frac{93}{86} - \frac{13}{10}$ >

Space for calculation

S No. 520

$\frac{11}{48} - \frac{35}{11}$ >

Space for calculation

Double Digit Different Denominator

Subtraction Fraction

Name: ____________________ **Date:** __________ **Score:** ________ 5

Start Time: ____________ **End Time:** ____________

S No. 521

$\frac{20}{18} - \frac{52}{49} >$

Space for calculation

S No. 522

$\frac{79}{19} - \frac{38}{57} >$

Space for calculation

S No. 523

$\frac{71}{20} - \frac{33}{97} >$

Space for calculation

S No. 524

$\frac{36}{53} - \frac{49}{70} >$

Space for calculation

S No. 525

$\frac{27}{72} - \frac{76}{43} >$

Space for calculation

S No. 526

$\frac{51}{10} - \frac{61}{68} >$

Space for calculation

S No. 527

$\frac{34}{60} - \frac{76}{99} >$

Space for calculation

S No. 528

$\frac{77}{34} - \frac{54}{61} >$

Space for calculation

S No. 529

$\frac{76}{84} - \frac{99}{53} >$

Space for calculation

S No. 530

$\frac{77}{59} - \frac{28}{70} >$

Space for calculation

Double Digit
Different
Denominator
Subtraction
Fraction

Name: ______________________ **Date:** ____________ **Score:** ____________
5

Start Time: ______________ **End Time:** ______________

S No. 531

$\frac{58}{19} - \frac{78}{94} >$

Space for calculation

S No. 536

$\frac{67}{94} - \frac{68}{71} >$

Space for calculation

S No. 532

$\frac{89}{19} - \frac{51}{49} >$

Space for calculation

S No. 537

$\frac{47}{63} - \frac{19}{74} >$

Space for calculation

S No. 533

$\frac{98}{81} - \frac{86}{97} >$

Space for calculation

S No. 538

$\frac{59}{16} - \frac{32}{89} >$

Space for calculation

S No. 534

$\frac{12}{63} - \frac{31}{81} >$

Space for calculation

S No. 539

$\frac{84}{29} - \frac{34}{32} >$

Space for calculation

S No. 535

$\frac{34}{48} - \frac{57}{11} >$

Space for calculation

S No. 540

$\frac{31}{89} - \frac{67}{79} >$

Space for calculation

Double Digit Different Denominator

Subtraction Fraction

Name: ____________ **Date:** ________ **Score:** ________ 5

Start Time: ________ **End Time:** ________

S No. 541

Space for calculation

$$\frac{39}{57} - \frac{79}{87} >$$

S No. 546

Space for calculation

$$\frac{82}{45} - \frac{66}{55} >$$

S No. 542

Space for calculation

$$\frac{38}{87} - \frac{70}{48} >$$

S No. 547

Space for calculation

$$\frac{66}{55} - \frac{92}{96} >$$

S No. 543

Space for calculation

$$\frac{13}{91} - \frac{99}{98} >$$

S No. 548

Space for calculation

$$\frac{50}{70} - \frac{37}{51} >$$

S No. 544

Space for calculation

$$\frac{24}{22} - \frac{36}{67} >$$

S No. 549

Space for calculation

$$\frac{57}{69} - \frac{73}{55} >$$

S No. 545

Space for calculation

$$\frac{38}{31} - \frac{36}{33} >$$

S No. 550

Space for calculation

$$\frac{13}{51} - \frac{76}{58} >$$

Double Digit Different Denominator

Subtraction Fraction

Name: ______ **Date:** ______ **Score:** ______

5

Start Time: ______ **End Time:** ______

S No. 551

$\frac{56}{30} - \frac{16}{65} >$

Space for calculation

S No. 556

$\frac{79}{32} - \frac{62}{48} >$

Space for calculation

S No. 552

$\frac{94}{36} - \frac{78}{62} >$

Space for calculation

S No. 557

$\frac{23}{29} - \frac{49}{15} >$

Space for calculation

S No. 553

$\frac{77}{11} - \frac{72}{11} >$

Space for calculation

S No. 558

$\frac{95}{39} - \frac{73}{65} >$

Space for calculation

S No. 554

$\frac{37}{92} - \frac{37}{48} >$

Space for calculation

S No. 559

$\frac{60}{53} - \frac{60}{95} >$

Space for calculation

S No. 555

$\frac{71}{81} - \frac{21}{99} >$

Space for calculation

S No. 560

$\frac{17}{89} - \frac{56}{90} >$

Space for calculation

Double Digit Different Denominator

Subtraction Fraction

Name: ____________ **Date:** ____________ **Score:** ____________ 5

Start Time: ____________ **End Time:** ____________

S No. 561

$\frac{30}{39} - \frac{48}{22} >$

Space for calculation

S No. 566

$\frac{85}{37} - \frac{14}{23} >$

Space for calculation

S No. 562

$\frac{63}{78} - \frac{73}{66} >$

Space for calculation

S No. 567

$\frac{43}{22} - \frac{81}{94} >$

Space for calculation

S No. 563

$\frac{59}{77} - \frac{77}{36} >$

Space for calculation

S No. 568

$\frac{51}{85} - \frac{85}{97} >$

Space for calculation

S No. 564

$\frac{83}{52} - \frac{17}{97} >$

Space for calculation

S No. 569

$\frac{92}{72} - \frac{94}{88} >$

Space for calculation

S No. 565

$\frac{28}{57} - \frac{95}{74} >$

Space for calculation

S No. 570

$\frac{64}{65} - \frac{67}{17} >$

Space for calculation

Double Digit Different Denominator

Subtraction Fraction

Name: ____________ Date: ________ Score: ______ 5

Start Time: ________ End Time: ________

S No. 571

Space for calculation

$\frac{60}{41} - \frac{64}{66}$ >

S No. 576

Space for calculation

$\frac{48}{40} - \frac{29}{28}$ >

S No. 572

Space for calculation

$\frac{32}{27} - \frac{66}{34}$ >

S No. 577

Space for calculation

$\frac{20}{78} - \frac{64}{12}$ >

S No. 573

Space for calculation

$\frac{57}{31} - \frac{50}{60}$ >

S No. 578

Space for calculation

$\frac{41}{30} - \frac{11}{19}$ >

S No. 574

Space for calculation

$\frac{51}{81} - \frac{99}{45}$ >

S No. 579

Space for calculation

$\frac{42}{37} - \frac{68}{63}$ >

S No. 575

Space for calculation

$\frac{40}{44} - \frac{16}{59}$ >

S No. 580

Space for calculation

$\frac{60}{77} - \frac{60}{86}$ >

Double Digit Different Denominator

Subtraction Fraction

Name: ______________________ **Date:** __________ **Score:** ________ 5

Start Time: ____________ **End Time:** ____________

S No. 581

$\frac{57}{54} - \frac{11}{14} >$

Space for calculation

S No. 582

$\frac{30}{43} - \frac{68}{37} >$

Space for calculation

S No. 583

$\frac{59}{21} - \frac{15}{18} >$

Space for calculation

S No. 584

$\frac{59}{51} - \frac{98}{29} >$

Space for calculation

S No. 585

$\frac{91}{61} - \frac{57}{95} >$

Space for calculation

S No. 586

$\frac{14}{46} - \frac{12}{96} >$

Space for calculation

S No. 587

$\frac{17}{75} - \frac{76}{87} >$

Space for calculation

S No. 588

$\frac{66}{45} - \frac{70}{53} >$

Space for calculation

S No. 589

$\frac{62}{99} - \frac{98}{45} >$

Space for calculation

S No. 590

$\frac{52}{24} - \frac{40}{97} >$

Space for calculation

Double Digit Different Denominator

Subtraction Fraction

Name: ____________ **Date:** ________ **Score:** ________

5

Start Time: ____________ **End Time:** ____________

S No. 591

$\frac{26}{58} - \frac{83}{18} >$

Space for calculation

S No. 596

$\frac{20}{67} - \frac{44}{14} >$

Space for calculation

S No. 592

$\frac{72}{87} - \frac{13}{35} >$

Space for calculation

S No. 597

$\frac{54}{33} - \frac{81}{72} >$

Space for calculation

S No. 593

$\frac{56}{43} - \frac{13}{10} >$

Space for calculation

S No. 598

$\frac{40}{71} - \frac{70}{84} >$

Space for calculation

S No. 594

$\frac{32}{72} - \frac{47}{49} >$

Space for calculation

S No. 599

$\frac{34}{29} - \frac{94}{49} >$

Space for calculation

S No. 595

$\frac{98}{62} - \frac{52}{19} >$

Space for calculation

S No. 600

$\frac{93}{59} - \frac{94}{90} >$

Space for calculation

Double Digit Different Denominator

Subtraction Fraction

Name: ____________ **Date:** ________ **Score:** ________ 5

Start Time: ________ **End Time:** ________

S No. 601

$\frac{85}{58} - \frac{64}{71} >$

Space for calculation

S No. 602

$\frac{46}{86} - \frac{77}{89} >$

Space for calculation

S No. 603

$\frac{90}{71} - \frac{56}{44} >$

Space for calculation

S No. 604

$\frac{75}{38} - \frac{93}{19} >$

Space for calculation

S No. 605

$\frac{95}{27} - \frac{35}{97} >$

Space for calculation

S No. 606

$\frac{24}{49} - \frac{11}{19} >$

Space for calculation

S No. 607

$\frac{17}{19} - \frac{98}{82} >$

Space for calculation

S No. 608

$\frac{82}{60} - \frac{13}{49} >$

Space for calculation

S No. 609

$\frac{94}{80} - \frac{89}{61} >$

Space for calculation

S No. 610

$\frac{29}{15} - \frac{36}{12} >$

Space for calculation

Double Digit Different Denominator

Subtraction Fraction

Name: ______________ **Date:** __________ **Score:** __________ 5

Start Time: __________ **End Time:** __________

S No. 611

$\frac{36}{32} - \frac{43}{10}$ >

Space for calculation

S No. 612

$\frac{78}{14} - \frac{76}{54}$ >

Space for calculation

S No. 613

$\frac{25}{65} - \frac{31}{81}$ >

Space for calculation

S No. 614

$\frac{81}{34} - \frac{34}{73}$ >

Space for calculation

S No. 615

$\frac{13}{68} - \frac{72}{40}$ >

Space for calculation

S No. 616

$\frac{85}{84} - \frac{44}{95}$ >

Space for calculation

S No. 617

$\frac{98}{57} - \frac{54}{77}$ >

Space for calculation

S No. 618

$\frac{73}{52} - \frac{83}{89}$ >

Space for calculation

S No. 619

$\frac{69}{34} - \frac{73}{24}$ >

Space for calculation

S No. 620

$\frac{18}{43} - \frac{48}{65}$ >

Space for calculation

Double Digit Different Denominator

Subtraction Fraction

Name: ________ **Date:** ________ **Score:** ________

5

Start Time: ________ **End Time:** ________

S No. 621

$\frac{49}{69} - \frac{36}{94}$ >

Space for calculation

S No. 626

$\frac{77}{20} - \frac{13}{67}$ >

Space for calculation

S No. 622

$\frac{27}{39} - \frac{21}{45}$ >

Space for calculation

S No. 627

$\frac{92}{25} - \frac{94}{87}$ >

Space for calculation

S No. 623

$\frac{74}{75} - \frac{87}{50}$ >

Space for calculation

S No. 628

$\frac{37}{50} - \frac{25}{77}$ >

Space for calculation

S No. 624

$\frac{62}{51} - \frac{44}{25}$ >

Space for calculation

S No. 629

$\frac{65}{88} - \frac{41}{99}$ >

Space for calculation

S No. 625

$\frac{26}{17} - \frac{42}{35}$ >

Space for calculation

S No. 630

$\frac{88}{89} - \frac{15}{87}$ >

Space for calculation

Double Digit Different Denominator

Subtraction Fraction

Name: ____________ **Date:** ____________ **Score:** ____________ 5

Start Time: ____________ **End Time:** ____________

S No. 631

Space for calculation

$\frac{49}{45} - \frac{98}{43} >$

S No. 632

Space for calculation

$\frac{76}{78} - \frac{68}{71} >$

S No. 633

Space for calculation

$\frac{93}{35} - \frac{45}{27} >$

S No. 634

Space for calculation

$\frac{87}{29} - \frac{98}{75} >$

S No. 635

Space for calculation

$\frac{75}{79} - \frac{38}{96} >$

S No. 636

Space for calculation

$\frac{98}{75} - \frac{77}{82} >$

S No. 637

Space for calculation

$\frac{39}{28} - \frac{33}{93} >$

S No. 638

Space for calculation

$\frac{72}{63} - \frac{46}{68} >$

S No. 639

Space for calculation

$\frac{59}{60} - \frac{86}{84} >$

S No. 640

Space for calculation

$\frac{50}{86} - \frac{87}{90} >$

Double Digit Different Denominator

Subtraction Fraction

Name: ______________________ **Date:** __________ **Score:** __________

5

Start Time: ______________ **End Time:** ______________

S No. 641

$\frac{96}{90} - \frac{24}{22} >$

Space for calculation

S No. 646

$\frac{76}{52} - \frac{28}{94} >$

Space for calculation

S No. 642

$\frac{77}{10} - \frac{43}{35} >$

Space for calculation

S No. 647

$\frac{55}{69} - \frac{59}{27} >$

Space for calculation

S No. 643

$\frac{59}{49} - \frac{70}{56} >$

Space for calculation

S No. 648

$\frac{69}{12} - \frac{15}{93} >$

Space for calculation

S No. 644

$\frac{37}{22} - \frac{49}{32} >$

Space for calculation

S No. 649

$\frac{67}{10} - \frac{82}{36} >$

Space for calculation

S No. 645

$\frac{69}{97} - \frac{37}{35} >$

Space for calculation

S No. 650

$\frac{16}{18} - \frac{86}{40} >$

Space for calculation

Double Digit Different Denominator

Subtraction Fraction

Name: ______________ Date: __________ Score: __________ 5

Start Time: __________ End Time: __________

S No. 651

Space for calculation

$\frac{57}{41} - \frac{19}{24} >$

S No. 656

Space for calculation

$\frac{90}{81} - \frac{78}{64} >$

S No. 652

Space for calculation

$\frac{12}{71} - \frac{47}{12} >$

S No. 657

Space for calculation

$\frac{88}{28} - \frac{14}{11} >$

S No. 653

Space for calculation

$\frac{58}{17} - \frac{10}{26} >$

S No. 658

Space for calculation

$\frac{80}{43} - \frac{91}{81} >$

S No. 654

Space for calculation

$\frac{93}{76} - \frac{95}{11} >$

S No. 659

Space for calculation

$\frac{45}{25} - \frac{77}{68} >$

S No. 655

Space for calculation

$\frac{49}{83} - \frac{70}{82} >$

S No. 660

Space for calculation

$\frac{77}{43} - \frac{92}{89} >$

Double Digit Different Denominator Subtraction Fraction

Name: ____________ **Date:** ________ **Score:** ________ 5

Start Time: ________ **End Time:** ________

S No. 661

$$\frac{24}{42} - \frac{97}{86} >$$

Space for calculation

S No. 666

$$\frac{14}{49} - \frac{27}{33} >$$

Space for calculation

S No. 662

$$\frac{49}{31} - \frac{97}{29} >$$

Space for calculation

S No. 667

$$\frac{96}{34} - \frac{99}{46} >$$

Space for calculation

S No. 663

$$\frac{37}{60} - \frac{61}{52} >$$

Space for calculation

S No. 668

$$\frac{92}{13} - \frac{45}{55} >$$

Space for calculation

S No. 664

$$\frac{97}{90} - \frac{93}{16} >$$

Space for calculation

S No. 669

$$\frac{57}{36} - \frac{33}{32} >$$

Space for calculation

S No. 665

$$\frac{75}{86} - \frac{51}{48} >$$

Space for calculation

S No. 670

$$\frac{87}{82} - \frac{26}{53} >$$

Space for calculation

Double Digit Different Denominator

Subtraction Fraction

Name: ____________ **Date:** ________ **Score:** ________

5

Start Time: ________ **End Time:** ________

S No. 671

Space for calculation

$\frac{73}{79} - \frac{87}{52} >$

S No. 672

Space for calculation

$\frac{29}{82} - \frac{65}{84} >$

S No. 673

Space for calculation

$\frac{46}{46} - \frac{26}{85} >$

S No. 674

Space for calculation

$\frac{43}{25} - \frac{72}{49} >$

S No. 675

Space for calculation

$\frac{38}{91} - \frac{32}{19} >$

S No. 676

Space for calculation

$\frac{11}{87} - \frac{31}{56} >$

S No. 677

Space for calculation

$\frac{37}{92} - \frac{87}{12} >$

S No. 678

Space for calculation

$\frac{13}{57} - \frac{59}{35} >$

S No. 679

Space for calculation

$\frac{21}{11} - \frac{18}{80} >$

S No. 680

Space for calculation

$\frac{53}{97} - \frac{50}{92} >$

Double Digit Different Denominator

Subtraction Fraction

Name: ______________________ **Date:** ____________ **Score:** __________ 5

Start Time: ______________ **End Time:** ______________

S No. 681

$\frac{63}{95} - \frac{73}{34} >$

Space for calculation

S No. 686

$\frac{62}{20} - \frac{83}{91} >$

Space for calculation

S No. 682

$\frac{27}{24} - \frac{56}{94} >$

Space for calculation

S No. 687

$\frac{25}{28} - \frac{43}{97} >$

Space for calculation

S No. 683

$\frac{17}{23} - \frac{69}{83} >$

Space for calculation

S No. 688

$\frac{34}{98} - \frac{86}{67} >$

Space for calculation

S No. 684

$\frac{86}{52} - \frac{73}{93} >$

Space for calculation

S No. 689

$\frac{26}{36} - \frac{72}{18} >$

Space for calculation

S No. 685

$\frac{35}{22} - \frac{71}{33} >$

Space for calculation

S No. 690

$\frac{39}{32} - \frac{34}{94} >$

Space for calculation

Double Digit Different Denominator Subtraction Fraction

Name: ______ Date: ______ Score: ______ 5

Start Time: ______ End Time: ______

S No. 691

Space for calculation

$\frac{92}{82} - \frac{97}{97}$ >

S No. 696

Space for calculation

$\frac{45}{90} - \frac{34}{25}$ >

S No. 692

Space for calculation

$\frac{43}{97} - \frac{45}{98}$ >

S No. 697

Space for calculation

$\frac{85}{31} - \frac{13}{78}$ >

S No. 693

Space for calculation

$\frac{68}{21} - \frac{26}{26}$ >

S No. 698

Space for calculation

$\frac{58}{85} - \frac{60}{10}$ >

S No. 694

Space for calculation

$\frac{81}{88} - \frac{38}{69}$ >

S No. 699

Space for calculation

$\frac{93}{13} - \frac{25}{12}$ >

S No. 695

Space for calculation

$\frac{13}{26} - \frac{80}{14}$ >

S No. 700

Space for calculation

$\frac{55}{15} - \frac{17}{25}$ >

S No. 701

$\frac{5}{9} \times \frac{9}{9}$ >

Space for calculation

S No. 702

$\frac{6}{5} \times \frac{7}{5}$ >

Space for calculation

S No. 703

$\frac{4}{3} \times \frac{4}{3}$ >

Space for calculation

S No. 704

$\frac{9}{9} \times \frac{2}{9}$ >

Space for calculation

S No. 705

$\frac{7}{7} \times \frac{4}{7}$ >

Space for calculation

S No. 706

$\frac{4}{3} \times \frac{9}{3}$ >

Space for calculation

S No. 707

$\frac{4}{9} \times \frac{3}{9}$ >

Space for calculation

S No. 708

$\frac{5}{5} \times \frac{4}{5}$ >

Space for calculation

S No. 709

$\frac{8}{4} \times \frac{8}{4}$ >

Space for calculation

S No. 710

$\frac{7}{2} \times \frac{2}{2}$ >

Space for calculation

Single Digit

Multiplication Fraction

Name: ____________ **Date:** ____________ **Score:** ____________

5

Start Time: ____________ **End Time:** ____________

S No. 711

Space for calculation

$\frac{7}{3} \times \frac{1}{3}$ >

S No. 716

Space for calculation

$\frac{8}{9} \times \frac{7}{9}$ >

S No. 712

Space for calculation

$\frac{4}{2} \times \frac{1}{2}$ >

S No. 717

Space for calculation

$\frac{2}{4} \times \frac{2}{4}$ >

S No. 713

Space for calculation

$\frac{8}{2} \times \frac{3}{2}$ >

S No. 718

Space for calculation

$\frac{3}{7} \times \frac{7}{7}$ >

S No. 714

Space for calculation

$\frac{6}{1} \times \frac{5}{1}$ >

S No. 719

Space for calculation

$\frac{2}{5} \times \frac{2}{5}$ >

S No. 715

Space for calculation

$\frac{7}{7} \times \frac{5}{7}$ >

S No. 720

Space for calculation

$\frac{3}{1} \times \frac{4}{1}$ >

S No. 721

Space for calculation

$\frac{86}{43} \times \frac{42}{43}$ >

S No. 722

Space for calculation

$\frac{58}{38} \times \frac{59}{38}$ >

S No. 723

Space for calculation

$\frac{67}{17} \times \frac{88}{17}$ >

S No. 724

Space for calculation

$\frac{83}{66} \times \frac{96}{66}$ >

S No. 725

Space for calculation

$\frac{84}{63} \times \frac{14}{63}$ >

S No. 726

Space for calculation

$\frac{34}{83} \times \frac{63}{83}$ >

S No. 727

Space for calculation

$\frac{76}{48} \times \frac{45}{48}$ >

S No. 728

Space for calculation

$\frac{45}{24} \times \frac{89}{24}$ >

S No. 729

Space for calculation

$\frac{48}{94} \times \frac{50}{94}$ >

S No. 730

Space for calculation

$\frac{59}{39} \times \frac{64}{39}$ >

Double Digit

Multiplication Fraction

Name: ______________________ **Date:** ____________ **Score:** ________ 5

Start Time: ______________ **End Time:** ______________

S No. 731

Space for calculation

$\frac{41}{50} \times \frac{34}{50} >$

S No. 736

Space for calculation

$\frac{47}{56} \times \frac{71}{56} >$

S No. 732

Space for calculation

$\frac{98}{54} \times \frac{47}{54} >$

S No. 737

Space for calculation

$\frac{97}{44} \times \frac{65}{44} >$

S No. 733

Space for calculation

$\frac{65}{15} \times \frac{18}{15} >$

S No. 738

Space for calculation

$\frac{30}{10} \times \frac{19}{10} >$

S No. 734

Space for calculation

$\frac{57}{71} \times \frac{72}{71} >$

S No. 739

Space for calculation

$\frac{92}{69} \times \frac{15}{69} >$

S No. 735

Space for calculation

$\frac{94}{71} \times \frac{57}{71} >$

S No. 740

Space for calculation

$\frac{66}{14} \times \frac{91}{14} >$

Double Digit

Multiplication Fraction

Name: ______________ **Date:** __________ **Score:** __________

5

Start Time: __________ **End Time:** __________

S No. 741

$\frac{94}{91} \times \frac{67}{91} >$

Space for calculation

S No. 746

$\frac{23}{79} \times \frac{82}{79} >$

Space for calculation

S No. 742

$\frac{80}{10} \times \frac{49}{10} >$

Space for calculation

S No. 747

$\frac{64}{96} \times \frac{39}{96} >$

Space for calculation

S No. 743

$\frac{46}{95} \times \frac{39}{95} >$

Space for calculation

S No. 748

$\frac{24}{75} \times \frac{53}{75} >$

Space for calculation

S No. 744

$\frac{98}{64} \times \frac{74}{64} >$

Space for calculation

S No. 749

$\frac{19}{73} \times \frac{82}{73} >$

Space for calculation

S No. 745

$\frac{77}{11} \times \frac{97}{11} >$

Space for calculation

S No. 750

$\frac{49}{60} \times \frac{33}{60} >$

Space for calculation

Double Digit

Multiplication Fraction

Name: ____________ **Date:** ____________ **Score:** ____________

5

Start Time: ____________ **End Time:** ____________

S No. 751

Space for calculation

$\frac{54}{97} \times \frac{87}{97}$ >

S No. 756

Space for calculation

$\frac{66}{62} \times \frac{31}{62}$ >

S No. 752

Space for calculation

$\frac{26}{54} \times \frac{88}{54}$ >

S No. 757

Space for calculation

$\frac{93}{30} \times \frac{34}{30}$ >

S No. 753

Space for calculation

$\frac{27}{76} \times \frac{31}{76}$ >

S No. 758

Space for calculation

$\frac{84}{62} \times \frac{43}{62}$ >

S No. 754

Space for calculation

$\frac{18}{35} \times \frac{24}{35}$ >

S No. 759

Space for calculation

$\frac{83}{99} \times \frac{72}{99}$ >

S No. 755

Space for calculation

$\frac{69}{64} \times \frac{48}{64}$ >

S No. 760

Space for calculation

$\frac{94}{91} \times \frac{43}{91}$ >

Double Digit

Multiplication Fraction

Name: ________ **Date:** ________ **Score:** ________ / 5

Start Time: ________ **End Time:** ________

S No. 761

$\frac{26}{45} \times \frac{29}{45} >$

Space for calculation

S No. 766

$\frac{62}{17} \times \frac{34}{17} >$

Space for calculation

S No. 762

$\frac{39}{30} \times \frac{71}{30} >$

Space for calculation

S No. 767

$\frac{54}{40} \times \frac{87}{40} >$

Space for calculation

S No. 763

$\frac{35}{56} \times \frac{23}{56} >$

Space for calculation

S No. 768

$\frac{83}{39} \times \frac{65}{39} >$

Space for calculation

S No. 764

$\frac{62}{57} \times \frac{95}{57} >$

Space for calculation

S No. 769

$\frac{35}{63} \times \frac{58}{63} >$

Space for calculation

S No. 765

$\frac{92}{10} \times \frac{83}{10} >$

Space for calculation

S No. 770

$\frac{76}{40} \times \frac{98}{40} >$

Space for calculation

Double Digit

Multiplication Fraction

Name: ______________ **Date:** ______________ **Score:** ______________ 5

Start Time: ______________ **End Time:** ______________

S No. 771

Space for calculation

$\frac{32}{71} \times \frac{68}{71} >$

S No. 776

Space for calculation

$\frac{12}{81} \times \frac{78}{81} >$

S No. 772

Space for calculation

$\frac{44}{34} \times \frac{89}{34} >$

S No. 777

Space for calculation

$\frac{43}{94} \times \frac{70}{94} >$

S No. 773

Space for calculation

$\frac{45}{87} \times \frac{43}{87} >$

S No. 778

Space for calculation

$\frac{52}{10} \times \frac{57}{10} >$

S No. 774

Space for calculation

$\frac{47}{81} \times \frac{54}{81} >$

S No. 779

Space for calculation

$\frac{70}{13} \times \frac{73}{13} >$

S No. 775

Space for calculation

$\frac{47}{55} \times \frac{40}{55} >$

S No. 780

Space for calculation

$\frac{99}{26} \times \frac{12}{26} >$

S No. 781

Space for calculation

$$\frac{4}{1} \times \frac{8}{5} >$$

S No. 782

Space for calculation

$$\frac{9}{4} \times \frac{6}{1} >$$

S No. 783

Space for calculation

$$\frac{9}{8} \times \frac{6}{9} >$$

S No. 784

Space for calculation

$$\frac{9}{1} \times \frac{1}{4} >$$

S No. 785

Space for calculation

$$\frac{3}{5} \times \frac{5}{2} >$$

S No. 786

Space for calculation

$$\frac{9}{8} \times \frac{1}{7} >$$

S No. 787

Space for calculation

$$\frac{6}{3} \times \frac{8}{9} >$$

S No. 788

Space for calculation

$$\frac{5}{6} \times \frac{7}{4} >$$

S No. 789

Space for calculation

$$\frac{1}{1} \times \frac{7}{7} >$$

S No. 790

Space for calculation

$$\frac{2}{2} \times \frac{2}{1} >$$

Single Digit Different Denominator

Multiplication Fraction

Name: ____________ **Date:** ________ **Score:** ________ 5

Start Time: ____________ **End Time:** ____________

S No. 791

Space for calculation

$$\frac{3}{4} \times \frac{2}{3} >$$

S No. 792

Space for calculation

$$\frac{5}{2} \times \frac{6}{9} >$$

S No. 793

Space for calculation

$$\frac{3}{7} \times \frac{4}{1} >$$

S No. 794

Space for calculation

$$\frac{1}{8} \times \frac{7}{7} >$$

S No. 795

Space for calculation

$$\frac{8}{5} \times \frac{4}{3} >$$

S No. 796

Space for calculation

$$\frac{1}{8} \times \frac{3}{7} >$$

S No. 797

Space for calculation

$$\frac{2}{1} \times \frac{1}{8} >$$

S No. 798

Space for calculation

$$\frac{4}{2} \times \frac{6}{7} >$$

S No. 799

Space for calculation

$$\frac{6}{1} \times \frac{7}{5} >$$

S No. 800

Space for calculation

$$\frac{9}{6} \times \frac{4}{3} >$$

S No. 801

Space for calculation

$\frac{87}{83} \times \frac{79}{39}$ >

S No. 802

Space for calculation

$\frac{61}{74} \times \frac{76}{55}$ >

S No. 803

Space for calculation

$\frac{48}{80} \times \frac{58}{60}$ >

S No. 804

Space for calculation

$\frac{89}{33} \times \frac{80}{70}$ >

S No. 805

Space for calculation

$\frac{95}{68} \times \frac{45}{51}$ >

S No. 806

Space for calculation

$\frac{86}{55} \times \frac{73}{72}$ >

S No. 807

Space for calculation

$\frac{51}{43} \times \frac{78}{28}$ >

S No. 808

Space for calculation

$\frac{92}{80} \times \frac{46}{36}$ >

S No. 809

Space for calculation

$\frac{80}{30} \times \frac{29}{18}$ >

S No. 810

Space for calculation

$\frac{59}{62} \times \frac{16}{11}$ >

Double Digit Different Denominator

Multiplication Fraction

Name: ____________ **Date:** ________ **Score:** ________ 5

Start Time: ________ **End Time:** ________

S No. 811

Space for calculation

$\frac{15}{44} \times \frac{43}{10}$ >

S No. 816

Space for calculation

$\frac{49}{77} \times \frac{47}{17}$ >

S No. 812

Space for calculation

$\frac{60}{28} \times \frac{46}{23}$ >

S No. 817

Space for calculation

$\frac{87}{18} \times \frac{82}{40}$ >

S No. 813

Space for calculation

$\frac{60}{66} \times \frac{68}{28}$ >

S No. 818

Space for calculation

$\frac{69}{46} \times \frac{49}{45}$ >

S No. 814

Space for calculation

$\frac{47}{51} \times \frac{72}{37}$ >

S No. 819

Space for calculation

$\frac{93}{15} \times \frac{12}{78}$ >

S No. 815

Space for calculation

$\frac{72}{84} \times \frac{65}{22}$ >

S No. 820

Space for calculation

$\frac{65}{54} \times \frac{61}{58}$ >

Double Digit Different Denominator

Multiplication Fraction

Name: ____________ **Date:** ________ **Score:** ________
5

Start Time: ____________ **End Time:** ____________

S No. 821

Space for calculation

$\frac{17}{78} \times \frac{91}{90}$ >

S No. 826

Space for calculation

$\frac{90}{88} \times \frac{62}{40}$ >

S No. 822

Space for calculation

$\frac{61}{60} \times \frac{25}{47}$ >

S No. 827

Space for calculation

$\frac{57}{41} \times \frac{64}{86}$ >

S No. 823

Space for calculation

$\frac{47}{24} \times \frac{89}{63}$ >

S No. 828

Space for calculation

$\frac{76}{52} \times \frac{33}{66}$ >

S No. 824

Space for calculation

$\frac{66}{95} \times \frac{30}{59}$ >

S No. 829

Space for calculation

$\frac{55}{99} \times \frac{98}{29}$ >

S No. 825

Space for calculation

$\frac{62}{55} \times \frac{17}{82}$ >

S No. 830

Space for calculation

$\frac{47}{77} \times \frac{25}{66}$ >

Double Digit Different Denominator Multiplication Fraction

Name: ______________ **Date:** __________ **Score:** __________
5

Start Time: __________ **End Time:** __________

S No. 831

Space for calculation

$\frac{51}{61} \times \frac{61}{75} >$

S No. 836

Space for calculation

$\frac{51}{30} \times \frac{84}{63} >$

S No. 832

Space for calculation

$\frac{27}{46} \times \frac{82}{24} >$

S No. 837

Space for calculation

$\frac{27}{56} \times \frac{23}{70} >$

S No. 833

Space for calculation

$\frac{29}{45} \times \frac{80}{16} >$

S No. 838

Space for calculation

$\frac{75}{39} \times \frac{49}{55} >$

S No. 834

Space for calculation

$\frac{58}{32} \times \frac{58}{36} >$

S No. 839

Space for calculation

$\frac{17}{87} \times \frac{50}{87} >$

S No. 835

Space for calculation

$\frac{57}{91} \times \frac{64}{86} >$

S No. 840

Space for calculation

$\frac{67}{11} \times \frac{21}{81} >$

Double Digit Different Denominator

Multiplication Fraction

Name: ______ **Date:** ______ **Score:** ______ 5

Start Time: ______ **End Time:** ______

S No. 841

Space for calculation

$\frac{32}{31} \times \frac{24}{15} >$

S No. 846

Space for calculation

$\frac{95}{43} \times \frac{21}{37} >$

S No. 842

Space for calculation

$\frac{31}{50} \times \frac{74}{33} >$

S No. 847

Space for calculation

$\frac{16}{26} \times \frac{68}{38} >$

S No. 843

Space for calculation

$\frac{29}{30} \times \frac{22}{70} >$

S No. 848

Space for calculation

$\frac{96}{14} \times \frac{26}{21} >$

S No. 844

Space for calculation

$\frac{29}{10} \times \frac{95}{23} >$

S No. 849

Space for calculation

$\frac{61}{51} \times \frac{23}{17} >$

S No. 845

Space for calculation

$\frac{26}{40} \times \frac{70}{21} >$

S No. 850

Space for calculation

$\frac{17}{87} \times \frac{39}{79} >$

Double Digit Different Denominator Multiplication Fraction

Name: ______ **Date:** ______ **Score:** ______ 5

Start Time: ______ **End Time:** ______

S No. 851

Space for calculation

$$\frac{59}{83} \times \frac{10}{94} >$$

S No. 852

Space for calculation

$$\frac{74}{71} \times \frac{43}{72} >$$

S No. 853

Space for calculation

$$\frac{72}{37} \times \frac{64}{78} >$$

S No. 854

Space for calculation

$$\frac{59}{54} \times \frac{17}{13} >$$

S No. 855

Space for calculation

$$\frac{42}{99} \times \frac{43}{97} >$$

S No. 856

Space for calculation

$$\frac{51}{31} \times \frac{45}{25} >$$

S No. 857

Space for calculation

$$\frac{20}{22} \times \frac{84}{79} >$$

S No. 858

Space for calculation

$$\frac{98}{88} \times \frac{74}{60} >$$

S No. 859

Space for calculation

$$\frac{53}{83} \times \frac{95}{49} >$$

S No. 860

Space for calculation

$$\frac{22}{40} \times \frac{49}{94} >$$

Double Digit Different Denominator

Multiplication Fraction

Name: ______ **Date:** ______ **Score:** ______ 5

Start Time: ______ **End Time:** ______

S No. 861

Space for calculation

$\frac{77}{34} \times \frac{97}{18} >$

S No. 866

Space for calculation

$\frac{89}{82} \times \frac{88}{27} >$

S No. 862

Space for calculation

$\frac{74}{56} \times \frac{17}{85} >$

S No. 867

Space for calculation

$\frac{91}{17} \times \frac{81}{38} >$

S No. 863

Space for calculation

$\frac{28}{85} \times \frac{79}{15} >$

S No. 868

Space for calculation

$\frac{26}{57} \times \frac{63}{48} >$

S No. 864

Space for calculation

$\frac{78}{68} \times \frac{61}{84} >$

S No. 869

Space for calculation

$\frac{10}{93} \times \frac{97}{88} >$

S No. 865

Space for calculation

$\frac{93}{86} \times \frac{13}{10} >$

S No. 870

Space for calculation

$\frac{11}{48} \times \frac{35}{11} >$

Double Digit Different Denominator

Multiplication Fraction

Name: ______________ **Date:** __________ **Score:** __________ 5

Start Time: __________ **End Time:** __________

S No. 871

$\frac{20}{18} \times \frac{52}{49}$ >

Space for calculation

S No. 872

$\frac{79}{19} \times \frac{38}{57}$ >

Space for calculation

S No. 873

$\frac{71}{20} \times \frac{33}{97}$ >

Space for calculation

S No. 874

$\frac{36}{53} \times \frac{49}{70}$ >

Space for calculation

S No. 875

$\frac{27}{72} \times \frac{76}{43}$ >

Space for calculation

S No. 876

$\frac{51}{10} \times \frac{61}{68}$ >

Space for calculation

S No. 877

$\frac{34}{60} \times \frac{76}{99}$ >

Space for calculation

S No. 878

$\frac{77}{34} \times \frac{54}{61}$ >

Space for calculation

S No. 879

$\frac{76}{84} \times \frac{99}{53}$ >

Space for calculation

S No. 880

$\frac{77}{59} \times \frac{28}{70}$ >

Space for calculation

Double Digit Different Denominator

Multiplication Fraction

Name: ____________________ **Date:** __________ **Score:** __________ 5

Start Time: ______________ **End Time:** ______________

S No. 881

$\frac{58}{19} \times \frac{78}{94} >$

Space for calculation

S No. 882

$\frac{89}{19} \times \frac{51}{49} >$

Space for calculation

S No. 883

$\frac{98}{81} \times \frac{86}{97} >$

Space for calculation

S No. 884

$\frac{12}{63} \times \frac{31}{81} >$

Space for calculation

S No. 885

$\frac{34}{48} \times \frac{57}{11} >$

Space for calculation

S No. 886

$\frac{67}{94} \times \frac{68}{71} >$

Space for calculation

S No. 887

$\frac{47}{63} \times \frac{19}{74} >$

Space for calculation

S No. 888

$\frac{59}{16} \times \frac{32}{89} >$

Space for calculation

S No. 889

$\frac{84}{29} \times \frac{34}{32} >$

Space for calculation

S No. 890

$\frac{31}{89} \times \frac{67}{79} >$

Space for calculation

Double Digit Different Denominator

Multiplication Fraction

Name: ____________________ **Date:** __________ **Score:** __________ 5

Start Time: __________ **End Time:** __________

S No. 891

Space for calculation

$\frac{39}{57} \times \frac{79}{87} >$

S No. 896

Space for calculation

$\frac{82}{45} \times \frac{66}{55} >$

S No. 892

Space for calculation

$\frac{38}{87} \times \frac{70}{48} >$

S No. 897

Space for calculation

$\frac{66}{55} \times \frac{92}{96} >$

S No. 893

Space for calculation

$\frac{13}{91} \times \frac{99}{98} >$

S No. 898

Space for calculation

$\frac{50}{70} \times \frac{37}{51} >$

S No. 894

Space for calculation

$\frac{24}{22} \times \frac{36}{67} >$

S No. 899

Space for calculation

$\frac{57}{69} \times \frac{73}{55} >$

S No. 895

Space for calculation

$\frac{38}{31} \times \frac{36}{33} >$

S No. 900

Space for calculation

$\frac{13}{51} \times \frac{76}{58} >$

Double Digit Different Denominator

Multiplication Fraction

Name: ______________________ Date: __________ Score: __________

5

Start Time: ______________ End Time: ______________

S No. 901

Space for calculation

$\frac{56}{30} \times \frac{16}{65} >$

S No. 906

Space for calculation

$\frac{79}{32} \times \frac{62}{48} >$

S No. 902

Space for calculation

$\frac{94}{36} \times \frac{78}{62} >$

S No. 907

Space for calculation

$\frac{23}{29} \times \frac{49}{15} >$

S No. 903

Space for calculation

$\frac{77}{11} \times \frac{72}{11} >$

S No. 908

Space for calculation

$\frac{95}{39} \times \frac{73}{65} >$

S No. 904

Space for calculation

$\frac{37}{92} \times \frac{37}{48} >$

S No. 909

Space for calculation

$\frac{60}{53} \times \frac{60}{95} >$

S No. 905

Space for calculation

$\frac{71}{81} \times \frac{21}{99} >$

S No. 910

Space for calculation

$\frac{17}{89} \times \frac{56}{90} >$

Double Digit Different Denominator Multiplication Fraction

Name: ______ Date: ______ Score: ______

5

Start Time: ______ End Time: ______

S No. 911

$\frac{30}{39} \times \frac{48}{22} >$

Space for calculation

S No. 916

$\frac{85}{37} \times \frac{14}{23} >$

Space for calculation

S No. 912

$\frac{63}{78} \times \frac{73}{66} >$

Space for calculation

S No. 917

$\frac{43}{22} \times \frac{81}{94} >$

Space for calculation

S No. 913

$\frac{59}{77} \times \frac{77}{36} >$

Space for calculation

S No. 918

$\frac{51}{85} \times \frac{85}{97} >$

Space for calculation

S No. 914

$\frac{83}{52} \times \frac{17}{97} >$

Space for calculation

S No. 919

$\frac{92}{72} \times \frac{94}{88} >$

Space for calculation

S No. 915

$\frac{28}{57} \times \frac{95}{74} >$

Space for calculation

S No. 920

$\frac{64}{65} \times \frac{67}{17} >$

Space for calculation

Double Digit Different Denominator

Multiplication Fraction

Name: ____________ Date: ________ Score: ______

5

Start Time: ________ End Time: ________

S No. 921

$$\frac{60}{41} \times \frac{64}{66}$$

Space for calculation

S No. 926

$$\frac{48}{40} \times \frac{29}{28}$$

Space for calculation

S No. 922

$$\frac{32}{27} \times \frac{66}{34}$$

Space for calculation

S No. 927

$$\frac{20}{78} \times \frac{64}{12}$$

Space for calculation

S No. 923

$$\frac{57}{31} \times \frac{50}{60}$$

Space for calculation

S No. 928

$$\frac{41}{30} \times \frac{11}{19}$$

Space for calculation

S No. 924

$$\frac{51}{81} \times \frac{99}{45}$$

Space for calculation

S No. 929

$$\frac{42}{37} \times \frac{68}{63}$$

Space for calculation

S No. 925

$$\frac{40}{44} \times \frac{16}{59}$$

Space for calculation

S No. 930

$$\frac{60}{77} \times \frac{60}{86}$$

Space for calculation

Double Digit Different Denominator

Multiplication Fraction

Name: ______ Date: ______ Score: ______ 5

Start Time: ______ End Time: ______

S No. 931

Space for calculation

$\frac{57}{54} \times \frac{11}{14}$ >

S No. 936

Space for calculation

$\frac{14}{46} \times \frac{12}{96}$ >

S No. 932

Space for calculation

$\frac{30}{43} \times \frac{68}{37}$ >

S No. 937

Space for calculation

$\frac{17}{75} \times \frac{76}{87}$ >

S No. 933

Space for calculation

$\frac{59}{21} \times \frac{15}{18}$ >

S No. 938

Space for calculation

$\frac{66}{45} \times \frac{70}{53}$ >

S No. 934

Space for calculation

$\frac{59}{51} \times \frac{98}{29}$ >

S No. 939

Space for calculation

$\frac{62}{99} \times \frac{98}{45}$ >

S No. 935

Space for calculation

$\frac{91}{61} \times \frac{57}{95}$ >

S No. 940

Space for calculation

$\frac{52}{24} \times \frac{40}{97}$ >

Double Digit Different Denominator

Multiplication Fraction

Name: ____________ **Date:** ________ **Score:** ______ 5

Start Time: ________ **End Time:** ________

S No. 941

$\frac{26}{58} \times \frac{83}{18}$ >

Space for calculation

S No. 946

$\frac{20}{67} \times \frac{44}{14}$ >

Space for calculation

S No. 942

$\frac{72}{87} \times \frac{13}{35}$ >

Space for calculation

S No. 947

$\frac{54}{33} \times \frac{81}{72}$ >

Space for calculation

S No. 943

$\frac{56}{43} \times \frac{13}{10}$ >

Space for calculation

S No. 948

$\frac{40}{71} \times \frac{70}{84}$ >

Space for calculation

S No. 944

$\frac{32}{72} \times \frac{47}{49}$ >

Space for calculation

S No. 949

$\frac{34}{29} \times \frac{94}{49}$ >

Space for calculation

S No. 945

$\frac{98}{62} \times \frac{52}{19}$ >

Space for calculation

S No. 950

$\frac{93}{59} \times \frac{94}{90}$ >

Space for calculation

Double Digit Different Denominator

Multiplication Fraction

Name: ______________________ **Date:** __________ **Score:** __________ 5

Start Time: ______________ **End Time:** ______________

S No. 951

Space for calculation

$$\frac{85}{58} \times \frac{64}{71} >$$

S No. 956

Space for calculation

$$\frac{24}{49} \times \frac{11}{19} >$$

S No. 952

Space for calculation

$$\frac{46}{86} \times \frac{77}{89} >$$

S No. 957

Space for calculation

$$\frac{17}{19} \times \frac{98}{82} >$$

S No. 953

Space for calculation

$$\frac{90}{71} \times \frac{56}{44} >$$

S No. 958

Space for calculation

$$\frac{82}{60} \times \frac{13}{49} >$$

S No. 954

Space for calculation

$$\frac{75}{38} \times \frac{93}{19} >$$

S No. 959

Space for calculation

$$\frac{94}{80} \times \frac{89}{61} >$$

S No. 955

Space for calculation

$$\frac{95}{27} \times \frac{35}{97} >$$

S No. 960

Space for calculation

$$\frac{29}{15} \times \frac{36}{12} >$$

Double Digit Different Denominator

Multiplication Fraction

Name: ____________ **Date:** ____________ **Score:** ____________ 5

Start Time: ____________ **End Time:** ____________

S No. 961

Space for calculation

$\frac{36}{32} \times \frac{43}{10} >$

S No. 966

Space for calculation

$\frac{85}{84} \times \frac{44}{95} >$

S No. 962

Space for calculation

$\frac{78}{14} \times \frac{76}{54} >$

S No. 967

Space for calculation

$\frac{98}{57} \times \frac{54}{77} >$

S No. 963

Space for calculation

$\frac{25}{65} \times \frac{31}{81} >$

S No. 968

Space for calculation

$\frac{73}{52} \times \frac{83}{89} >$

S No. 964

Space for calculation

$\frac{81}{34} \times \frac{34}{73} >$

S No. 969

Space for calculation

$\frac{69}{34} \times \frac{73}{24} >$

S No. 965

Space for calculation

$\frac{13}{68} \times \frac{72}{40} >$

S No. 970

Space for calculation

$\frac{18}{43} \times \frac{48}{65} >$

Double Digit Different Denominator

Multiplication Fraction

Name: ____________ **Date:** ____________ **Score:** ____________ 5

Start Time: ____________ **End Time:** ____________

S No. 971

Space for calculation

$\frac{49}{69} \times \frac{36}{94} >$

S No. 976

Space for calculation

$\frac{77}{20} \times \frac{13}{67} >$

S No. 972

Space for calculation

$\frac{27}{39} \times \frac{21}{45} >$

S No. 977

Space for calculation

$\frac{92}{25} \times \frac{94}{87} >$

S No. 973

Space for calculation

$\frac{74}{75} \times \frac{87}{50} >$

S No. 978

Space for calculation

$\frac{37}{50} \times \frac{25}{77} >$

S No. 974

Space for calculation

$\frac{62}{51} \times \frac{44}{25} >$

S No. 979

Space for calculation

$\frac{65}{88} \times \frac{41}{99} >$

S No. 975

Space for calculation

$\frac{26}{17} \times \frac{42}{35} >$

S No. 980

Space for calculation

$\frac{88}{89} \times \frac{15}{87} >$

Double Digit Different Denominator

Multiplication Fraction

Name: ____________ **Date:** ________ **Score:** ________ 5

Start Time: ________ **End Time:** ________

S No. 981

$$\frac{49}{45} \times \frac{98}{43}$$

Space for calculation

S No. 986

$$\frac{98}{75} \times \frac{77}{82}$$

Space for calculation

S No. 982

$$\frac{76}{78} \times \frac{68}{71}$$

Space for calculation

S No. 987

$$\frac{39}{28} \times \frac{33}{93}$$

Space for calculation

S No. 983

$$\frac{93}{35} \times \frac{45}{27}$$

Space for calculation

S No. 988

$$\frac{72}{63} \times \frac{46}{68}$$

Space for calculation

S No. 984

$$\frac{87}{29} \times \frac{98}{75}$$

Space for calculation

S No. 989

$$\frac{59}{60} \times \frac{86}{84}$$

Space for calculation

S No. 985

$$\frac{75}{79} \times \frac{38}{96}$$

Space for calculation

S No. 990

$$\frac{50}{86} \times \frac{87}{90}$$

Space for calculation

Double Digit Different Denominator

Multiplication Fraction

Name: ____________ **Date:** ________ **Score:** ________ 5

Start Time: ________ **End Time:** ________

S No. 991

Space for calculation

$$\frac{96}{90} \times \frac{24}{22} >$$

S No. 996

Space for calculation

$$\frac{76}{52} \times \frac{28}{94} >$$

S No. 992

Space for calculation

$$\frac{77}{10} \times \frac{43}{35} >$$

S No. 997

Space for calculation

$$\frac{55}{69} \times \frac{59}{27} >$$

S No. 993

Space for calculation

$$\frac{59}{49} \times \frac{70}{56} >$$

S No. 998

Space for calculation

$$\frac{69}{12} \times \frac{15}{93} >$$

S No. 994

Space for calculation

$$\frac{37}{22} \times \frac{49}{32} >$$

S No. 999

Space for calculation

$$\frac{67}{10} \times \frac{82}{36} >$$

S No. 995

Space for calculation

$$\frac{69}{97} \times \frac{37}{35} >$$

S No. 1000

Space for calculation

$$\frac{16}{18} \times \frac{86}{40} >$$

Double Digit Different Denominator

Multiplication Fraction

Name: ____________ **Date:** ________ **Score:** ________

5

Start Time: ________ **End Time:** ________

S No. 1001

Space for calculation

$\frac{57}{41} \times \frac{19}{24} >$

S No. 1006

Space for calculation

$\frac{90}{81} \times \frac{78}{64} >$

S No. 1002

Space for calculation

$\frac{12}{71} \times \frac{47}{12} >$

S No. 1007

Space for calculation

$\frac{88}{28} \times \frac{14}{11} >$

S No. 1003

Space for calculation

$\frac{58}{17} \times \frac{10}{26} >$

S No. 1008

Space for calculation

$\frac{80}{43} \times \frac{91}{81} >$

S No. 1004

Space for calculation

$\frac{93}{76} \times \frac{95}{11} >$

S No. 1009

Space for calculation

$\frac{45}{25} \times \frac{77}{68} >$

S No. 1005

Space for calculation

$\frac{49}{83} \times \frac{70}{82} >$

S No. 1010

Space for calculation

$\frac{77}{43} \times \frac{92}{89} >$

Double Digit Different Denominator Multiplication Fraction

Name: ____________ **Date:** ________ **Score:** ________ 5

Start Time: ____________ **End Time:** ____________

S No. 1011

Space for calculation

$$\frac{24}{42} \times \frac{97}{86} >$$

S No. 1016

Space for calculation

$$\frac{14}{49} \times \frac{27}{33} >$$

S No. 1012

Space for calculation

$$\frac{49}{31} \times \frac{97}{29} >$$

S No. 1017

Space for calculation

$$\frac{96}{34} \times \frac{99}{46} >$$

S No. 1013

Space for calculation

$$\frac{37}{60} \times \frac{61}{52} >$$

S No. 1018

Space for calculation

$$\frac{92}{13} \times \frac{45}{55} >$$

S No. 1014

Space for calculation

$$\frac{97}{90} \times \frac{93}{16} >$$

S No. 1019

Space for calculation

$$\frac{57}{36} \times \frac{33}{32} >$$

S No. 1015

Space for calculation

$$\frac{75}{86} \times \frac{51}{48} >$$

S No. 1020

Space for calculation

$$\frac{87}{82} \times \frac{26}{53} >$$

Double Digit Different Denominator

Multiplication Fraction

Name: ______________ **Date:** __________ **Score:** __________ 5

Start Time: __________ **End Time:** __________

S No. 1021

$\frac{73}{79} \times \frac{87}{52}$ >

Space for calculation

S No. 1026

$\frac{11}{87} \times \frac{31}{56}$ >

Space for calculation

S No. 1022

$\frac{29}{82} \times \frac{65}{84}$ >

Space for calculation

S No. 1027

$\frac{37}{92} \times \frac{87}{12}$ >

Space for calculation

S No. 1023

$\frac{46}{46} \times \frac{26}{85}$ >

Space for calculation

S No. 1028

$\frac{13}{57} \times \frac{59}{35}$ >

Space for calculation

S No. 1024

$\frac{43}{25} \times \frac{72}{49}$ >

Space for calculation

S No. 1029

$\frac{21}{11} \times \frac{18}{80}$ >

Space for calculation

S No. 1025

$\frac{38}{91} \times \frac{32}{19}$ >

Space for calculation

S No. 1030

$\frac{53}{97} \times \frac{50}{92}$ >

Space for calculation

Double Digit Different Denominator Multiplication Fraction

Name: ____________ Date: ____________ Score: ____________ 5

Start Time: ____________ End Time: ____________

S No. 1031

Space for calculation

$\frac{63}{95} \times \frac{73}{34}$ >

S No. 1036

Space for calculation

$\frac{62}{20} \times \frac{83}{91}$ >

S No. 1032

Space for calculation

$\frac{27}{24} \times \frac{56}{94}$ >

S No. 1037

Space for calculation

$\frac{25}{28} \times \frac{43}{97}$ >

S No. 1033

Space for calculation

$\frac{17}{23} \times \frac{69}{83}$ >

S No. 1038

Space for calculation

$\frac{34}{98} \times \frac{86}{67}$ >

S No. 1034

Space for calculation

$\frac{86}{52} \times \frac{73}{93}$ >

S No. 1039

Space for calculation

$\frac{26}{36} \times \frac{72}{18}$ >

S No. 1035

Space for calculation

$\frac{35}{22} \times \frac{71}{33}$ >

S No. 1040

Space for calculation

$\frac{39}{32} \times \frac{34}{94}$ >

Double Digit Different Denominator

Multiplication Fraction

Name: ____________ **Date:** ____________ **Score:** ____________ 5

Start Time: ____________ **End Time:** ____________

S No. 1041

$\frac{92}{82} \times \frac{97}{97}$ >

Space for calculation

S No. 1046

$\frac{45}{90} \times \frac{34}{25}$ >

Space for calculation

S No. 1042

$\frac{43}{97} \times \frac{45}{98}$ >

Space for calculation

S No. 1047

$\frac{85}{31} \times \frac{13}{78}$ >

Space for calculation

S No. 1043

$\frac{68}{21} \times \frac{26}{26}$ >

Space for calculation

S No. 1048

$\frac{58}{85} \times \frac{60}{10}$ >

Space for calculation

S No. 1044

$\frac{81}{88} \times \frac{38}{69}$ >

Space for calculation

S No. 1049

$\frac{93}{13} \times \frac{25}{12}$ >

Space for calculation

S No. 1045

$\frac{13}{26} \times \frac{80}{14}$ >

Space for calculation

S No. 1050

$\frac{55}{15} \times \frac{17}{25}$ >

Space for calculation

Single Digit

Division Fraction

Name: ____________ **Date:** ________ **Score:** ________

5

Start Time: ________ **End Time:** ________

S No. 1051

Space for calculation

$\frac{5}{9} \div \frac{9}{9}$ >

S No. 1052

Space for calculation

$\frac{6}{5} \div \frac{7}{5}$ >

S No. 1053

Space for calculation

$\frac{4}{3} \div \frac{4}{3}$ >

S No. 1054

Space for calculation

$\frac{9}{9} \div \frac{2}{9}$ >

S No. 1055

Space for calculation

$\frac{7}{7} \div \frac{4}{7}$ >

S No. 1056

Space for calculation

$\frac{4}{3} \div \frac{9}{3}$ >

S No. 1057

Space for calculation

$\frac{4}{9} \div \frac{3}{9}$ >

S No. 1058

Space for calculation

$\frac{5}{5} \div \frac{4}{5}$ >

S No. 1059

Space for calculation

$\frac{8}{4} \div \frac{8}{4}$ >

S No. 1060

Space for calculation

$\frac{7}{2} \div \frac{2}{2}$ >

Single Digit

Division Fraction

Name: ______ **Date:** ______ **Score:** ______ 5

Start Time: ______ **End Time:** ______

S No. 1061

$\frac{7}{3} \div \frac{1}{3}$ >

Space for calculation

S No. 1066

$\frac{8}{9} \div \frac{7}{9}$ >

Space for calculation

S No. 1062

$\frac{4}{2} \div \frac{1}{2}$ >

Space for calculation

S No. 1067

$\frac{2}{4} \div \frac{2}{4}$ >

Space for calculation

S No. 1063

$\frac{8}{2} \div \frac{3}{2}$ >

Space for calculation

S No. 1068

$\frac{3}{7} \div \frac{7}{7}$ >

Space for calculation

S No. 1064

$\frac{6}{1} \div \frac{5}{1}$ >

Space for calculation

S No. 1069

$\frac{2}{5} \div \frac{2}{5}$ >

Space for calculation

S No. 1065

$\frac{7}{7} \div \frac{5}{7}$ >

Space for calculation

S No. 1070

$\frac{3}{1} \div \frac{4}{1}$ >

Space for calculation

Double Digit

Division Fraction

Name: ____________ **Date:** ____________ **Score:** ____________

5

Start Time: ____________ **End Time:** ____________

S No. 1071

$$\frac{86}{43} \div \frac{42}{43} >$$

Space for calculation

S No. 1076

$$\frac{34}{83} \div \frac{63}{83} >$$

Space for calculation

S No. 1072

$$\frac{58}{38} \div \frac{59}{38} >$$

Space for calculation

S No. 1077

$$\frac{76}{48} \div \frac{45}{48} >$$

Space for calculation

S No. 1073

$$\frac{67}{17} \div \frac{88}{17} >$$

Space for calculation

S No. 1078

$$\frac{45}{24} \div \frac{89}{24} >$$

Space for calculation

S No. 1074

$$\frac{83}{66} \div \frac{96}{66} >$$

Space for calculation

S No. 1079

$$\frac{48}{94} \div \frac{50}{94} >$$

Space for calculation

S No. 1075

$$\frac{84}{63} \div \frac{14}{63} >$$

Space for calculation

S No. 1080

$$\frac{59}{39} \div \frac{64}{39} >$$

Space for calculation

Double Digit

Division Fraction

Name: ____________ **Date:** ____________ **Score:** ____________ 5

Start Time: ____________ **End Time:** ____________

S No. 1081

$\frac{41}{50} \div \frac{34}{50} >$

Space for calculation

S No. 1082

$\frac{98}{54} \div \frac{47}{54} >$

Space for calculation

S No. 1083

$\frac{65}{15} \div \frac{18}{15} >$

Space for calculation

S No. 1084

$\frac{57}{71} \div \frac{72}{71} >$

Space for calculation

S No. 1085

$\frac{94}{71} \div \frac{57}{71} >$

Space for calculation

S No. 1086

$\frac{47}{56} \div \frac{71}{56} >$

Space for calculation

S No. 1087

$\frac{97}{44} \div \frac{65}{44} >$

Space for calculation

S No. 1088

$\frac{30}{10} \div \frac{19}{10} >$

Space for calculation

S No. 1089

$\frac{92}{69} \div \frac{15}{69} >$

Space for calculation

S No. 1090

$\frac{66}{14} \div \frac{91}{14} >$

Space for calculation

Double Digit

Division Fraction

Name: ______ **Date:** ______ **Score:** ______

5

Start Time: ______ **End Time:** ______

S No. 1091

$\frac{94}{91} \div \frac{67}{91} >$

Space for calculation

S No. 1096

$\frac{23}{79} \div \frac{82}{79} >$

Space for calculation

S No. 1092

$\frac{80}{10} \div \frac{49}{10} >$

Space for calculation

S No. 1097

$\frac{64}{96} \div \frac{39}{96} >$

Space for calculation

S No. 1093

$\frac{46}{95} \div \frac{39}{95} >$

Space for calculation

S No. 1098

$\frac{24}{75} \div \frac{53}{75} >$

Space for calculation

S No. 1094

$\frac{98}{64} \div \frac{74}{64} >$

Space for calculation

S No. 1099

$\frac{19}{73} \div \frac{82}{73} >$

Space for calculation

S No. 1095

$\frac{77}{11} \div \frac{97}{11} >$

Space for calculation

S No. 1100

$\frac{49}{60} \div \frac{33}{60} >$

Space for calculation

Double Digit

Division Fraction

Name: ____________ **Date:** ____________ **Score:** ____________ 5

Start Time: ____________ **End Time:** ____________

S No. 1101

$\frac{54}{97} \div \frac{87}{97} >$

Space for calculation

S No. 1106

$\frac{66}{62} \div \frac{31}{62} >$

Space for calculation

S No. 1102

$\frac{26}{54} \div \frac{88}{54} >$

Space for calculation

S No. 1107

$\frac{93}{30} \div \frac{34}{30} >$

Space for calculation

S No. 1103

$\frac{27}{76} \div \frac{31}{76} >$

Space for calculation

S No. 1108

$\frac{84}{62} \div \frac{43}{62} >$

Space for calculation

S No. 1104

$\frac{18}{35} \div \frac{24}{35} >$

Space for calculation

S No. 1109

$\frac{83}{99} \div \frac{72}{99} >$

Space for calculation

S No. 1105

$\frac{69}{64} \div \frac{48}{64} >$

Space for calculation

S No. 1110

$\frac{94}{91} \div \frac{43}{91} >$

Space for calculation

Double Digit

Division Fraction

Name: ______________________ **Date:** ____________ **Score:** ________ 5

Start Time: ______________ **End Time:** ______________

S No. 1111

Space for calculation

$\frac{26}{45} \div \frac{29}{45}$ >

S No. 1112

Space for calculation

$\frac{39}{30} \div \frac{71}{30}$ >

S No. 1113

Space for calculation

$\frac{35}{56} \div \frac{23}{56}$ >

S No. 1114

Space for calculation

$\frac{62}{57} \div \frac{95}{57}$ >

S No. 1115

Space for calculation

$\frac{92}{10} \div \frac{83}{10}$ >

S No. 1116

Space for calculation

$\frac{62}{17} \div \frac{34}{17}$ >

S No. 1117

Space for calculation

$\frac{54}{40} \div \frac{87}{40}$ >

S No. 1118

Space for calculation

$\frac{83}{39} \div \frac{65}{39}$ >

S No. 1119

Space for calculation

$\frac{35}{63} \div \frac{58}{63}$ >

S No. 1120

Space for calculation

$\frac{76}{40} \div \frac{98}{40}$ >

Double Digit

Division Fraction

Name: ______________ **Date:** __________ **Score:** ________ 5

Start Time: ____________ **End Time:** ____________

S No. 1121

$\frac{32}{71} \div \frac{68}{71}$ >

Space for calculation

S No. 1126

$\frac{12}{81} \div \frac{78}{81}$ >

Space for calculation

S No. 1122

$\frac{44}{34} \div \frac{89}{34}$ >

Space for calculation

S No. 1127

$\frac{43}{94} \div \frac{70}{94}$ >

Space for calculation

S No. 1123

$\frac{45}{87} \div \frac{43}{87}$ >

Space for calculation

S No. 1128

$\frac{52}{10} \div \frac{57}{10}$ >

Space for calculation

S No. 1124

$\frac{47}{81} \div \frac{54}{81}$ >

Space for calculation

S No. 1129

$\frac{70}{13} \div \frac{73}{13}$ >

Space for calculation

S No. 1125

$\frac{47}{55} \div \frac{40}{55}$ >

Space for calculation

S No. 1130

$\frac{99}{26} \div \frac{12}{26}$ >

Space for calculation

Single Digit
Different
Denominator
Division
Fraction

Name: ____________ **Date:** ________ **Score:** ________
5

Start Time: ________ **End Time:** ________

S No. 1131

Space for calculation

$\frac{4}{1} \div \frac{8}{5}$ >

S No. 1132

Space for calculation

$\frac{9}{4} \div \frac{6}{1}$ >

S No. 1133

Space for calculation

$\frac{9}{8} \div \frac{6}{9}$ >

S No. 1134

Space for calculation

$\frac{9}{1} \div \frac{1}{4}$ >

S No. 1135

Space for calculation

$\frac{3}{5} \div \frac{5}{2}$ >

S No. 1136

Space for calculation

$\frac{9}{8} \div \frac{1}{7}$ >

S No. 1137

Space for calculation

$\frac{6}{3} \div \frac{8}{9}$ >

S No. 1138

Space for calculation

$\frac{5}{6} \div \frac{7}{4}$ >

S No. 1139

Space for calculation

$\frac{1}{1} \div \frac{7}{7}$ >

S No. 1140

Space for calculation

$\frac{2}{2} \div \frac{2}{1}$ >

Single Digit Different Denominator

Division Fraction

Name: ______________________ **Date:** __________ **Score:** __________ 5

Start Time: ______________ **End Time:** ______________

S No. 1141

$\frac{3}{4} \div \frac{2}{3}$ >

Space for calculation

S No. 1146

$\frac{1}{8} \div \frac{3}{7}$ >

Space for calculation

S No. 1142

$\frac{5}{2} \div \frac{6}{9}$ >

Space for calculation

S No. 1147

$\frac{2}{1} \div \frac{1}{8}$ >

Space for calculation

S No. 1143

$\frac{3}{7} \div \frac{4}{1}$ >

Space for calculation

S No. 1148

$\frac{4}{2} \div \frac{6}{7}$ >

Space for calculation

S No. 1144

$\frac{1}{8} \div \frac{7}{7}$ >

Space for calculation

S No. 1149

$\frac{6}{1} \div \frac{7}{5}$ >

Space for calculation

S No. 1145

$\frac{8}{5} \div \frac{4}{3}$ >

Space for calculation

S No. 1150

$\frac{9}{6} \div \frac{4}{3}$ >

Space for calculation

Double Digit Different Denominator

Division Fraction

Name: ____________ **Date:** ________ **Score:** ______ 5

Start Time: __________ **End Time:** __________

S No. 1151

Space for calculation

$\frac{87}{83} \div \frac{79}{39}$ >

S No. 1156

Space for calculation

$\frac{86}{55} \div \frac{73}{72}$ >

S No. 1152

Space for calculation

$\frac{61}{74} \div \frac{76}{55}$ >

S No. 1157

Space for calculation

$\frac{51}{43} \div \frac{78}{28}$ >

S No. 1153

Space for calculation

$\frac{48}{80} \div \frac{58}{60}$ >

S No. 1158

Space for calculation

$\frac{92}{80} \div \frac{46}{36}$ >

S No. 1154

Space for calculation

$\frac{89}{33} \div \frac{80}{70}$ >

S No. 1159

Space for calculation

$\frac{80}{30} \div \frac{29}{18}$ >

S No. 1155

Space for calculation

$\frac{95}{68} \div \frac{45}{51}$ >

S No. 1160

Space for calculation

$\frac{59}{62} \div \frac{16}{11}$ >

Double Digit Different Denominator Division Fraction

Name: ____________ Date: ____________ Score: ________ 5

Start Time: ____________ End Time: ____________

S No. 1161

$\frac{15}{44} \div \frac{43}{10} >$

Space for calculation

S No. 1166

$\frac{49}{77} \div \frac{47}{17} >$

Space for calculation

S No. 1162

$\frac{60}{28} \div \frac{46}{23} >$

Space for calculation

S No. 1167

$\frac{87}{18} \div \frac{82}{40} >$

Space for calculation

S No. 1163

$\frac{60}{66} \div \frac{68}{28} >$

Space for calculation

S No. 1168

$\frac{69}{46} \div \frac{49}{45} >$

Space for calculation

S No. 1164

$\frac{47}{51} \div \frac{72}{37} >$

Space for calculation

S No. 1169

$\frac{93}{15} \div \frac{12}{78} >$

Space for calculation

S No. 1165

$\frac{72}{84} \div \frac{65}{22} >$

Space for calculation

S No. 1170

$\frac{65}{54} \div \frac{61}{58} >$

Space for calculation

Double Digit Different Denominator

Division Fraction

Name: ______________ **Date:** __________ **Score:** __________

5

Start Time: __________ **End Time:** __________

S No. 1171

Space for calculation

$\frac{17}{78} \div \frac{91}{90}$ >

S No. 1176

Space for calculation

$\frac{90}{88} \div \frac{62}{40}$ >

S No. 1172

Space for calculation

$\frac{61}{60} \div \frac{25}{47}$ >

S No. 1177

Space for calculation

$\frac{57}{41} \div \frac{64}{86}$ >

S No. 1173

Space for calculation

$\frac{47}{24} \div \frac{89}{63}$ >

S No. 1178

Space for calculation

$\frac{76}{52} \div \frac{33}{66}$ >

S No. 1174

Space for calculation

$\frac{66}{95} \div \frac{30}{59}$ >

S No. 1179

Space for calculation

$\frac{55}{99} \div \frac{98}{29}$ >

S No. 1175

Space for calculation

$\frac{62}{55} \div \frac{17}{82}$ >

S No. 1180

Space for calculation

$\frac{47}{77} \div \frac{25}{66}$ >

Double Digit Different Denominator

Division Fraction

Name: ____________ **Date:** ________ **Score:** ____ / 5

Start Time: ________ **End Time:** ________

S No. 1181

$\frac{51}{61} \div \frac{61}{75}$ >

Space for calculation

S No. 1182

$\frac{27}{46} \div \frac{82}{24}$ >

Space for calculation

S No. 1183

$\frac{29}{45} \div \frac{80}{16}$ >

Space for calculation

S No. 1184

$\frac{58}{32} \div \frac{58}{36}$ >

Space for calculation

S No. 1185

$\frac{57}{91} \div \frac{64}{86}$ >

Space for calculation

S No. 1186

$\frac{51}{30} \div \frac{84}{63}$ >

Space for calculation

S No. 1187

$\frac{27}{56} \div \frac{23}{70}$ >

Space for calculation

S No. 1188

$\frac{75}{39} \div \frac{49}{55}$ >

Space for calculation

S No. 1189

$\frac{17}{87} \div \frac{50}{87}$ >

Space for calculation

S No. 1190

$\frac{67}{11} \div \frac{21}{81}$ >

Space for calculation

Double Digit Different Denominator

Division Fraction

Name: ____________ **Date:** ________ **Score:** ________ 5

Start Time: __________ **End Time:** __________

S No. 1191

Space for calculation

$$\frac{32}{31} \div \frac{24}{15} >$$

S No. 1196

Space for calculation

$$\frac{95}{43} \div \frac{21}{37} >$$

S No. 1192

Space for calculation

$$\frac{31}{50} \div \frac{74}{33} >$$

S No. 1197

Space for calculation

$$\frac{16}{26} \div \frac{68}{38} >$$

S No. 1193

Space for calculation

$$\frac{29}{30} \div \frac{22}{70} >$$

S No. 1198

Space for calculation

$$\frac{96}{14} \div \frac{26}{21} >$$

S No. 1194

Space for calculation

$$\frac{29}{10} \div \frac{95}{23} >$$

S No. 1199

Space for calculation

$$\frac{61}{51} \div \frac{23}{17} >$$

S No. 1195

Space for calculation

$$\frac{26}{40} \div \frac{70}{21} >$$

S No. 1200

Space for calculation

$$\frac{17}{87} \div \frac{39}{79} >$$

Double Digit Different Denominator

Division Fraction

Name: ______ **Date:** ______ **Score:** ______ 5

Start Time: ______ **End Time:** ______

S No. 1201

Space for calculation

$\frac{59}{83} \div \frac{10}{94}$ >

S No. 1206

Space for calculation

$\frac{51}{31} \div \frac{45}{25}$ >

S No. 1202

Space for calculation

$\frac{74}{71} \div \frac{43}{72}$ >

S No. 1207

Space for calculation

$\frac{20}{22} \div \frac{84}{79}$ >

S No. 1203

Space for calculation

$\frac{72}{37} \div \frac{64}{78}$ >

S No. 1208

Space for calculation

$\frac{98}{88} \div \frac{74}{60}$ >

S No. 1204

Space for calculation

$\frac{59}{54} \div \frac{17}{13}$ >

S No. 1209

Space for calculation

$\frac{53}{83} \div \frac{95}{49}$ >

S No. 1205

Space for calculation

$\frac{42}{99} \div \frac{43}{97}$ >

S No. 1210

Space for calculation

$\frac{22}{40} \div \frac{49}{94}$ >

Double Digit Different Denominator

Division Fraction

Name: ____________ **Date:** ____________ **Score:** ____________ 5

Start Time: ____________ **End Time:** ____________

S No. 1211

Space for calculation

$$\frac{77}{34} \div \frac{97}{18} >$$

S No. 1216

Space for calculation

$$\frac{89}{82} \div \frac{88}{27} >$$

S No. 1212

Space for calculation

$$\frac{74}{56} \div \frac{17}{85} >$$

S No. 1217

Space for calculation

$$\frac{91}{17} \div \frac{81}{38} >$$

S No. 1213

Space for calculation

$$\frac{28}{85} \div \frac{79}{15} >$$

S No. 1218

Space for calculation

$$\frac{26}{57} \div \frac{63}{48} >$$

S No. 1214

Space for calculation

$$\frac{78}{68} \div \frac{61}{84} >$$

S No. 1219

Space for calculation

$$\frac{10}{93} \div \frac{97}{88} >$$

S No. 1215

Space for calculation

$$\frac{93}{86} \div \frac{13}{10} >$$

S No. 1220

Space for calculation

$$\frac{11}{48} \div \frac{35}{11} >$$

Double Digit Different Denominator Division Fraction

Name: ____________ **Date:** ____________ **Score:** ______ 5

Start Time: ____________ **End Time:** ____________

S No. 1221

$$\frac{20}{18} \div \frac{52}{49}$$

Space for calculation

S No. 1226

$$\frac{51}{10} \div \frac{61}{68}$$

Space for calculation

S No. 1222

$$\frac{79}{19} \div \frac{38}{57}$$

Space for calculation

S No. 1227

$$\frac{34}{60} \div \frac{76}{99}$$

Space for calculation

S No. 1223

$$\frac{71}{20} \div \frac{33}{97}$$

Space for calculation

S No. 1228

$$\frac{77}{34} \div \frac{54}{61}$$

Space for calculation

S No. 1224

$$\frac{36}{53} \div \frac{49}{70}$$

Space for calculation

S No. 1229

$$\frac{76}{84} \div \frac{99}{53}$$

Space for calculation

S No. 1225

$$\frac{27}{72} \div \frac{76}{43}$$

Space for calculation

S No. 1230

$$\frac{77}{59} \div \frac{28}{70}$$

Space for calculation

Double Digit Different Denominator

Division Fraction

Name: ______________________ **Date:** __________ **Score:** __________

5

Start Time: ______________ **End Time:** ______________

S No. 1231

Space for calculation

$\frac{58}{19} \div \frac{78}{94}$ >

S No. 1236

Space for calculation

$\frac{67}{94} \div \frac{68}{71}$ >

S No. 1232

Space for calculation

$\frac{89}{19} \div \frac{51}{49}$ >

S No. 1237

Space for calculation

$\frac{47}{63} \div \frac{19}{74}$ >

S No. 1233

Space for calculation

$\frac{98}{81} \div \frac{86}{97}$ >

S No. 1238

Space for calculation

$\frac{59}{16} \div \frac{32}{89}$ >

S No. 1234

Space for calculation

$\frac{12}{63} \div \frac{31}{81}$ >

S No. 1239

Space for calculation

$\frac{84}{29} \div \frac{34}{32}$ >

S No. 1235

Space for calculation

$\frac{34}{48} \div \frac{57}{11}$ >

S No. 1240

Space for calculation

$\frac{31}{89} \div \frac{67}{79}$ >

Double Digit Different Denominator Division Fraction

Name: ______ **Date:** ______ **Score:** ______
5

Start Time: ______ **End Time:** ______

S No. 1241

$\frac{39}{57} \div \frac{79}{87}$

Space for calculation

S No. 1246

$\frac{82}{45} \div \frac{66}{55}$

Space for calculation

S No. 1242

$\frac{38}{87} \div \frac{70}{48}$

Space for calculation

S No. 1247

$\frac{66}{55} \div \frac{92}{96}$

Space for calculation

S No. 1243

$\frac{13}{91} \div \frac{99}{98}$

Space for calculation

S No. 1248

$\frac{50}{70} \div \frac{37}{51}$

Space for calculation

S No. 1244

$\frac{24}{22} \div \frac{36}{67}$

Space for calculation

S No. 1249

$\frac{57}{69} \div \frac{73}{55}$

Space for calculation

S No. 1245

$\frac{38}{31} \div \frac{36}{33}$

Space for calculation

S No. 1250

$\frac{13}{51} \div \frac{76}{58}$

Space for calculation

Double Digit Different Denominator

Division Fraction

Name: ______ **Date:** ______ **Score:** ______ 5

Start Time: ______ **End Time:** ______

S No. 1251

Space for calculation

$\frac{56}{30} \div \frac{16}{65}$ >

S No. 1256

Space for calculation

$\frac{79}{32} \div \frac{62}{48}$ >

S No. 1252

Space for calculation

$\frac{94}{36} \div \frac{78}{62}$ >

S No. 1257

Space for calculation

$\frac{23}{29} \div \frac{49}{15}$ >

S No. 1253

Space for calculation

$\frac{77}{11} \div \frac{72}{11}$ >

S No. 1258

Space for calculation

$\frac{95}{39} \div \frac{73}{65}$ >

S No. 1254

Space for calculation

$\frac{37}{92} \div \frac{37}{48}$ >

S No. 1259

Space for calculation

$\frac{60}{53} \div \frac{60}{95}$ >

S No. 1255

Space for calculation

$\frac{71}{81} \div \frac{21}{99}$ >

S No. 1260

Space for calculation

$\frac{17}{89} \div \frac{56}{90}$ >

Double Digit Different Denominator

Division Fraction

Name: ____________ **Date:** ____________ **Score:** ____________ 5

Start Time: ____________ **End Time:** ____________

S No. 1261

$\frac{30}{39} \div \frac{48}{22}$ >

Space for calculation

S No. 1266

$\frac{85}{37} \div \frac{14}{23}$ >

Space for calculation

S No. 1262

$\frac{63}{78} \div \frac{73}{66}$ >

Space for calculation

S No. 1267

$\frac{43}{22} \div \frac{81}{94}$ >

Space for calculation

S No. 1263

$\frac{59}{77} \div \frac{77}{36}$ >

Space for calculation

S No. 1268

$\frac{51}{85} \div \frac{85}{97}$ >

Space for calculation

S No. 1264

$\frac{83}{52} \div \frac{17}{97}$ >

Space for calculation

S No. 1269

$\frac{92}{72} \div \frac{94}{88}$ >

Space for calculation

S No. 1265

$\frac{28}{57} \div \frac{95}{74}$ >

Space for calculation

S No. 1270

$\frac{64}{65} \div \frac{67}{17}$ >

Space for calculation

Double Digit Different Denominator

Division Fraction

Name: ____________ **Date:** ________ **Score:** ________ 5

Start Time: ____________ **End Time:** ____________

S No. 1271

Space for calculation

$\frac{60}{41} \div \frac{64}{66} >$

S No. 1276

Space for calculation

$\frac{48}{40} \div \frac{29}{28} >$

S No. 1272

Space for calculation

$\frac{32}{27} \div \frac{66}{34} >$

S No. 1277

Space for calculation

$\frac{20}{78} \div \frac{64}{12} >$

S No. 1273

Space for calculation

$\frac{57}{31} \div \frac{50}{60} >$

S No. 1278

Space for calculation

$\frac{41}{30} \div \frac{11}{19} >$

S No. 1274

Space for calculation

$\frac{51}{81} \div \frac{99}{45} >$

S No. 1279

Space for calculation

$\frac{42}{37} \div \frac{68}{63} >$

S No. 1275

Space for calculation

$\frac{40}{44} \div \frac{16}{59} >$

S No. 1280

Space for calculation

$\frac{60}{77} \div \frac{60}{86} >$

Double Digit Different Denominator

Division Fraction

Name: ______________________ **Date:** __________ **Score:** __________ 5

Start Time: ______________ **End Time:** ______________

S No. 1281

$$\frac{57}{54} \div \frac{11}{14} >$$

Space for calculation

S No. 1286

$$\frac{14}{46} \div \frac{12}{96} >$$

Space for calculation

S No. 1282

$$\frac{30}{43} \div \frac{68}{37} >$$

Space for calculation

S No. 1287

$$\frac{17}{75} \div \frac{76}{87} >$$

Space for calculation

S No. 1283

$$\frac{59}{21} \div \frac{15}{18} >$$

Space for calculation

S No. 1288

$$\frac{66}{45} \div \frac{70}{53} >$$

Space for calculation

S No. 1284

$$\frac{59}{51} \div \frac{98}{29} >$$

Space for calculation

S No. 1289

$$\frac{62}{99} \div \frac{98}{45} >$$

Space for calculation

S No. 1285

$$\frac{91}{61} \div \frac{57}{95} >$$

Space for calculation

S No. 1290

$$\frac{52}{24} \div \frac{40}{97} >$$

Space for calculation

Double Digit Different Denominator

Division Fraction

Name: ______ Date: ______ Score: ______

5

Start Time: ______ End Time: ______

S No. 1291

Space for calculation

$\frac{26}{58} \div \frac{83}{18} >$

S No. 1296

Space for calculation

$\frac{20}{67} \div \frac{44}{14} >$

S No. 1292

Space for calculation

$\frac{72}{87} \div \frac{13}{35} >$

S No. 1297

Space for calculation

$\frac{54}{33} \div \frac{81}{72} >$

S No. 1293

Space for calculation

$\frac{56}{43} \div \frac{13}{10} >$

S No. 1298

Space for calculation

$\frac{40}{71} \div \frac{70}{84} >$

S No. 1294

Space for calculation

$\frac{32}{72} \div \frac{47}{49} >$

S No. 1299

Space for calculation

$\frac{34}{29} \div \frac{94}{49} >$

S No. 1295

Space for calculation

$\frac{98}{62} \div \frac{52}{19} >$

S No. 1300

Space for calculation

$\frac{93}{59} \div \frac{94}{90} >$

Double Digit Different Denominator

Division Fraction

Name: ____________ **Date:** ____________ **Score:** ____________ 5

Start Time: ____________ **End Time:** ____________

S No. 1301

$\frac{85}{58} \div \frac{64}{71}$ >

Space for calculation

S No. 1306

$\frac{24}{49} \div \frac{11}{19}$ >

Space for calculation

S No. 1302

$\frac{46}{86} \div \frac{77}{89}$ >

Space for calculation

S No. 1307

$\frac{17}{19} \div \frac{98}{82}$ >

Space for calculation

S No. 1303

$\frac{90}{71} \div \frac{56}{44}$ >

Space for calculation

S No. 1308

$\frac{82}{60} \div \frac{13}{49}$ >

Space for calculation

S No. 1304

$\frac{75}{38} \div \frac{93}{19}$ >

Space for calculation

S No. 1309

$\frac{94}{80} \div \frac{89}{61}$ >

Space for calculation

S No. 1305

$\frac{95}{27} \div \frac{35}{97}$ >

Space for calculation

S No. 1310

$\frac{29}{15} \div \frac{36}{12}$ >

Space for calculation

Double Digit Different Denominator

Division Fraction

Name: ______ **Date:** ______ **Score:** ______ 5

Start Time: ______ **End Time:** ______

S No. 1311

Space for calculation

$\frac{36}{32} \div \frac{43}{10}$ >

S No. 1312

Space for calculation

$\frac{78}{14} \div \frac{76}{54}$ >

S No. 1313

Space for calculation

$\frac{25}{65} \div \frac{31}{81}$ >

S No. 1314

Space for calculation

$\frac{81}{34} \div \frac{34}{73}$ >

S No. 1315

Space for calculation

$\frac{13}{68} \div \frac{72}{40}$ >

S No. 1316

Space for calculation

$\frac{85}{84} \div \frac{44}{95}$ >

S No. 1317

Space for calculation

$\frac{98}{57} \div \frac{54}{77}$ >

S No. 1318

Space for calculation

$\frac{73}{52} \div \frac{83}{89}$ >

S No. 1319

Space for calculation

$\frac{69}{34} \div \frac{73}{24}$ >

S No. 1320

Space for calculation

$\frac{18}{43} \div \frac{48}{65}$ >

Double Digit Different Denominator Division Fraction

Name: ____________ **Date:** ________ **Score:** ________ 5

Start Time: ________ **End Time:** ________

S No. 1321 — Space for calculation

$$\frac{49}{69} \div \frac{36}{94}$$

S No. 1326 — Space for calculation

$$\frac{77}{20} \div \frac{13}{67}$$

S No. 1322 — Space for calculation

$$\frac{27}{39} \div \frac{21}{45}$$

S No. 1327 — Space for calculation

$$\frac{92}{25} \div \frac{94}{87}$$

S No. 1323 — Space for calculation

$$\frac{74}{75} \div \frac{87}{50}$$

S No. 1328 — Space for calculation

$$\frac{37}{50} \div \frac{25}{77}$$

S No. 1324 — Space for calculation

$$\frac{62}{51} \div \frac{44}{25}$$

S No. 1329 — Space for calculation

$$\frac{65}{88} \div \frac{41}{99}$$

S No. 1325 — Space for calculation

$$\frac{26}{17} \div \frac{42}{35}$$

S No. 1330 — Space for calculation

$$\frac{88}{89} \div \frac{15}{87}$$

Double Digit Different Denominator

Division Fraction

Name: ____________ **Date:** ________ **Score:** ________ 5

Start Time: ________ **End Time:** ________

S No. 1331

Space for calculation

$$\frac{49}{45} \div \frac{98}{43} >$$

S No. 1336

Space for calculation

$$\frac{98}{75} \div \frac{77}{82} >$$

S No. 1332

Space for calculation

$$\frac{76}{78} \div \frac{68}{71} >$$

S No. 1337

Space for calculation

$$\frac{39}{28} \div \frac{33}{93} >$$

S No. 1333

Space for calculation

$$\frac{93}{35} \div \frac{45}{27} >$$

S No. 1338

Space for calculation

$$\frac{72}{63} \div \frac{46}{68} >$$

S No. 1334

Space for calculation

$$\frac{87}{29} \div \frac{98}{75} >$$

S No. 1339

Space for calculation

$$\frac{59}{60} \div \frac{86}{84} >$$

S No. 1335

Space for calculation

$$\frac{75}{79} \div \frac{38}{96} >$$

S No. 1340

Space for calculation

$$\frac{50}{86} \div \frac{87}{90} >$$

Double Digit Different Denominator

Division Fraction

Name: ____________ **Date:** ________ **Score:** ______ 5

Start Time: ________ **End Time:** ________

S No. 1341

Space for calculation

$\frac{96}{90} \div \frac{24}{22} >$

S No. 1346

Space for calculation

$\frac{76}{52} \div \frac{28}{94} >$

S No. 1342

Space for calculation

$\frac{77}{10} \div \frac{43}{35} >$

S No. 1347

Space for calculation

$\frac{55}{69} \div \frac{59}{27} >$

S No. 1343

Space for calculation

$\frac{59}{49} \div \frac{70}{56} >$

S No. 1348

Space for calculation

$\frac{69}{12} \div \frac{15}{93} >$

S No. 1344

Space for calculation

$\frac{37}{22} \div \frac{49}{32} >$

S No. 1349

Space for calculation

$\frac{67}{10} \div \frac{82}{36} >$

S No. 1345

Space for calculation

$\frac{69}{97} \div \frac{37}{35} >$

S No. 1350

Space for calculation

$\frac{16}{18} \div \frac{86}{40} >$

Double Digit Different Denominator

Division Fraction

Name: ____________ **Date:** ________ **Score:** ________ 5

Start Time: ________ **End Time:** ________

S No. 1351

Space for calculation

$$\frac{57}{41} \div \frac{19}{24} >$$

S No. 1356

Space for calculation

$$\frac{90}{81} \div \frac{78}{64} >$$

S No. 1352

Space for calculation

$$\frac{12}{71} \div \frac{47}{12} >$$

S No. 1357

Space for calculation

$$\frac{88}{28} \div \frac{14}{11} >$$

S No. 1353

Space for calculation

$$\frac{58}{17} \div \frac{10}{26} >$$

S No. 1358

Space for calculation

$$\frac{80}{43} \div \frac{91}{81} >$$

S No. 1354

Space for calculation

$$\frac{93}{76} \div \frac{95}{11} >$$

S No. 1359

Space for calculation

$$\frac{45}{25} \div \frac{77}{68} >$$

S No. 1355

Space for calculation

$$\frac{49}{83} \div \frac{70}{82} >$$

S No. 1360

Space for calculation

$$\frac{77}{43} \div \frac{92}{89} >$$

Double Digit Different Denominator

Division Fraction

Name: ______________________ **Date:** ____________ **Score:** ________ 5

Start Time: ______________ **End Time:** ______________

S No. 1361

Space for calculation

$\frac{24}{42} \div \frac{97}{86}$ >

S No. 1366

Space for calculation

$\frac{14}{49} \div \frac{27}{33}$ >

S No. 1362

Space for calculation

$\frac{49}{31} \div \frac{97}{29}$ >

S No. 1367

Space for calculation

$\frac{96}{34} \div \frac{99}{46}$ >

S No. 1363

Space for calculation

$\frac{37}{60} \div \frac{61}{52}$ >

S No. 1368

Space for calculation

$\frac{92}{13} \div \frac{45}{55}$ >

S No. 1364

Space for calculation

$\frac{97}{90} \div \frac{93}{16}$ >

S No. 1369

Space for calculation

$\frac{57}{36} \div \frac{33}{32}$ >

S No. 1365

Space for calculation

$\frac{75}{86} \div \frac{51}{48}$ >

S No. 1370

Space for calculation

$\frac{87}{82} \div \frac{26}{53}$ >

Double Digit Different Denominator

Division Fraction

Name: ____________ **Date:** ____________ **Score:** ____________ 5

Start Time: ____________ **End Time:** ____________

S No. 1371

$\frac{73}{79} \div \frac{87}{52}$ >

Space for calculation

S No. 1376

$\frac{11}{87} \div \frac{31}{56}$ >

Space for calculation

S No. 1372

$\frac{29}{82} \div \frac{65}{84}$ >

Space for calculation

S No. 1377

$\frac{37}{92} \div \frac{87}{12}$ >

Space for calculation

S No. 1373

$\frac{46}{46} \div \frac{26}{85}$ >

Space for calculation

S No. 1378

$\frac{13}{57} \div \frac{59}{35}$ >

Space for calculation

S No. 1374

$\frac{43}{25} \div \frac{72}{49}$ >

Space for calculation

S No. 1379

$\frac{21}{11} \div \frac{18}{80}$ >

Space for calculation

S No. 1375

$\frac{38}{91} \div \frac{32}{19}$ >

Space for calculation

S No. 1380

$\frac{53}{97} \div \frac{50}{92}$ >

Space for calculation

Double Digit Different Denominator

Division Fraction

Name: ______________ **Date:** __________ **Score:** __________

5

Start Time: __________ **End Time:** __________

S No. 1381

$\frac{63}{95} \div \frac{73}{34}$ >

Space for calculation

S No. 1386

$\frac{62}{20} \div \frac{83}{91}$ >

Space for calculation

S No. 1382

$\frac{27}{24} \div \frac{56}{94}$ >

Space for calculation

S No. 1387

$\frac{25}{28} \div \frac{43}{97}$ >

Space for calculation

S No. 1383

$\frac{17}{23} \div \frac{69}{83}$ >

Space for calculation

S No. 1388

$\frac{34}{98} \div \frac{86}{67}$ >

Space for calculation

S No. 1384

$\frac{86}{52} \div \frac{73}{93}$ >

Space for calculation

S No. 1389

$\frac{26}{36} \div \frac{72}{18}$ >

Space for calculation

S No. 1385

$\frac{35}{22} \div \frac{71}{33}$ >

Space for calculation

S No. 1390

$\frac{39}{32} \div \frac{34}{94}$ >

Space for calculation

Double Digit Different Denominator Division Fraction

Name: ______ **Date:** ______ **Score:** ______ 5

Start Time: ______ **End Time:** ______

S No. 1391

$\frac{92}{82} \div \frac{97}{97} >$

Space for calculation

S No. 1392

$\frac{43}{97} \div \frac{45}{98} >$

Space for calculation

S No. 1393

$\frac{68}{21} \div \frac{26}{26} >$

Space for calculation

S No. 1394

$\frac{81}{88} \div \frac{38}{69} >$

Space for calculation

S No. 1395

$\frac{13}{26} \div \frac{80}{14} >$

Space for calculation

S No. 1396

$\frac{45}{90} \div \frac{34}{25} >$

Space for calculation

S No. 1397

$\frac{85}{31} \div \frac{13}{78} >$

Space for calculation

S No. 1398

$\frac{58}{85} \div \frac{60}{10} >$

Space for calculation

S No. 1399

$\frac{93}{13} \div \frac{25}{12} >$

Space for calculation

S No. 1400

$\frac{55}{15} \div \frac{17}{25} >$

Space for calculation

Answers

Page No: 1 Sum: + *S. No: 1* **Ans: 14/9**	Page No: 1 Sum: + *S. No: 2* **Ans: 13/5**	Page No: 1 Sum: + *S. No: 3* **Ans: 8/3**	Page No: 1 Sum: + *S. No: 4* **Ans: 11/9**
Page No: 1 Sum: + *S. No: 5* **Ans: 11/7**	Page No: 1 Sum: + *S. No: 6* **Ans: 13/3**	Page No: 1 Sum: + *S. No: 7* **Ans: 7/9**	Page No: 1 Sum: + *S. No: 8* **Ans: 9/5**
Page No: 1 Sum: + *S. No: 9* **Ans: /4**	Page No: 1 Sum: + *S. No: 10* **Ans: 9/2**	Page No: 2 Sum: + *S. No: 11* **Ans: 8/3**	Page No: 2 Sum: + *S. No: 12* **Ans: 5/2**
Page No: 2 Sum: + *S. No: 13* **Ans: 11/2**	Page No: 2 Sum: + *S. No: 14* **Ans:11**	Page No: 2 Sum: + *S. No: 15* **Ans: 12/7**	Page No: 2 Sum: + *S. No: 16* **Ans: 5/3**
Page No: 2 Sum: + *S. No: 17* **Ans: /1**	Page No: 2 Sum: + *S. No: 18* **Ans: 10/7**	Page No: 2 Sum: + *S. No: 19* **Ans: 4/5**	Page No: 2 Sum: + *S. No: 20* **Ans: /7**
Page No: 3 Sum: + *S. No: 21* **Ans: 128/43**	Page No: 3 Sum: + *S. No: 22* **Ans: 117/38**	Page No: 3 Sum: + *S. No: 23* **Ans: 155/17**	Page No: 3 Sum: + *S. No: 24* **Ans: 179/66**
Page No: 3 Sum: + *S. No: 25* **Ans: 14/9**	Page No: 3 Sum: + *S. No: 26* **Ans: 97/83**	Page No: 3 Sum: + *S. No: 27* **Ans: 121/48**	Page No: 3 Sum: + *S. No: 28* **Ans: 67/12**

Answers

Page No: 3
Sum: +

S. No: 29
Ans: 49/47

Page No: 3
Sum: +

S. No: 30
Ans: 41/13

Page No: 4
Sum: +

S. No: 31
Ans: 3/2

Page No: 4
Sum: +

S. No: 32
Ans: 145/54

Page No: 4
Sum: +

S. No: 33
Ans: 83/15

Page No: 4
Sum: +

S. No: 34
Ans: 129/71

Page No: 4
Sum: +

S. No: 35
Ans: 151/71

Page No: 4
Sum: +

S. No: 36
Ans: 59/28

Page No: 4
Sum: +

S. No: 37
Ans: 81/22

Page No: 4
Sum: +

S. No: 38
Ans: 49/10

Page No: 4
Sum: +

S. No: 39
Ans: 107/69

Page No: 4
Sum: +

S. No: 40
Ans: 157/14

Page No: 5
Sum: +

S. No: 41
Ans: 23/13

Page No: 5
Sum: +

S. No: 42
Ans: 129/10

Page No: 5
Sum: +

S. No: 43
Ans: 17/19

Page No: 5
Sum: +

S. No: 44
Ans: 43/16

Page No: 5
Sum: +

S. No: 45
Ans: 174/11

Page No: 5
Sum: +

S. No: 46
Ans: 105/79

Page No: 5
Sum: +

S. No: 47
Ans: 103/96

Page No: 5
Sum: +

S. No: 48
Ans: 77/75

Page No: 5
Sum: +

S. No: 49
Ans: 101/73

Page No: 5
Sum: +

S. No: 50
Ans: 41/30

Page No: 6
Sum: +

S. No: 51
Ans: 141/97

Page No: 6
Sum: +

S. No: 52
Ans: 19/9

Page No: 6
Sum: +

S. No: 53
Ans: 29/38

Page No: 6
Sum: +

S. No: 54
Ans: 6/5

Page No: 6
Sum: +

S. No: 55
Ans: 117/64

Page No: 6
Sum: +

S. No: 56
Ans: 97/62

Answers

Page No: 6 Sum: + *S. No: 57* **Ans: 127/30**	Page No: 6 Sum: + *S. No: 58* **Ans: 127/62**	Page No: 6 Sum: + *S. No: 59* **Ans: 155/99**	Page No: 6 Sum: + *S. No: 60* **Ans: 137/91**
Page No: 7 Sum: + *S. No: 61* **Ans: 11/9**	Page No: 7 Sum: + *S. No: 62* **Ans: 11/3**	Page No: 7 Sum: + *S. No: 63* **Ans: 29/28**	Page No: 7 Sum: + *S. No: 64* **Ans: 157/57**
Page No: 7 Sum: + *S. No: 65* **Ans: 35/2**	Page No: 7 Sum: + *S. No: 66* **Ans: 96/17**	Page No: 7 Sum: + *S. No: 67* **Ans: 141/40**	Page No: 7 Sum: + *S. No: 68* **Ans: 148/39**
Page No: 7 Sum: + *S. No: 69* **Ans: 31/21**	Page No: 7 Sum: + *S. No: 70* **Ans: 87/20**	Page No: 8 Sum: + *S. No: 71* **Ans: 100/71**	Page No: 8 Sum: + *S. No: 72* **Ans: 133/34**
Page No: 8 Sum: + *S. No: 73* **Ans: 88/87**	Page No: 8 Sum: + *S. No: 74* **Ans: 101/81**	Page No: 8 Sum: + *S. No: 75* **Ans: 87/55**	Page No: 8 Sum: + *S. No: 76* **Ans: 10/9**
Page No: 8 Sum: + *S. No: 77* **Ans: 113/94**	Page No: 8 Sum: + *S. No: 78* **Ans: 109/10**	Page No: 8 Sum: + *S. No: 79* **Ans: 11**	Page No: 8 Sum: + *S. No: 80* **Ans: 111/26**
Page No: 9 Sum: + *S. No: 81* **Ans: 28/5**	Page No: 9 Sum: + *S. No: 82* **Ans: 33/4**	Page No: 9 Sum: + *S. No: 83* **Ans: 43/24**	Page No: 9 Sum: + *S. No: 84* **Ans: 37/4**

Answers

Page No: 9
Sum: +

S. No: 85
Ans: 31/10

Page No: 9
Sum: +

S. No: 86
Ans: 71/56

Page No: 9
Sum: +

S. No: 87
Ans: 26/9

Page No: 9
Sum: +

S. No: 88
Ans: 31/12

Page No: 9
Sum: +

S. No: 89
Ans: 2

Page No: 9
Sum: +

S. No: 90
Ans: 3

Page No: 10
Sum: +

S. No: 91
Ans: 17/12

Page No: 10
Sum: +

S. No: 92
Ans: 19/6

Page No: 10
Sum: +

S. No: 93
Ans: 31/7

Page No: 10
Sum: +

S. No: 94
Ans: 9/8

Page No: 10
Sum: +

S. No: 95
Ans: 44/15

Page No: 10
Sum: +

S. No: 96
Ans: 31/56

Page No: 10
Sum: +

S. No: 97
Ans: 17/8

Page No: 10
Sum: +

S. No: 98
Ans: 20/7

Page No: 10
Sum: +

S. No: 99
Ans: 37/5

Page No: 10
Sum: +

S. No: 100
Ans: 17/6

Page No: 11
Sum: +

S. No: 101
Ans: 9950/3237

Page No: 11
Sum: +

S. No: 102
Ans: 8979/4070

Page No: 11
Sum: +

S. No: 103
Ans: 47/30

Page No: 11
Sum: +

S. No: 104
Ans: 887/231

Page No: 11
Sum: +

S. No: 105
Ans: 155/68

Page No: 11
Sum: +

S. No: 106
Ans: 10207/3960

Page No: 11
Sum: +

S. No: 107
Ans: 2391/602

Page No: 11
Sum: +

S. No: 108
Ans: 437/180

Page No: 11
Sum: +

S. No: 109
Ans: 77/18

Page No: 11
Sum: +

S. No: 110
Ans: 1641/682

Page No: 12
Sum: +

S. No: 111
Ans: 1021/220

Page No: 12
Sum: +

S. No: 112
Ans: 29/7

Answers

Page No: 12
Sum: +

S. No: 113
Ans: 257/77

Page No: 12
Sum: +

S. No: 114
Ans: 5411/1887

Page No: 12
Sum: +

S. No: 115
Ans: 587/154

Page No: 12
Sum: +

S. No: 116
Ans: 636/187

Page No: 12
Sum: +

S. No: 117
Ans: 413/60

Page No: 12
Sum: +

S. No: 118
Ans: 233/90

Page No: 12
Sum: +

S. No: 119
Ans: 413/65

Page No: 12
Sum: +

S. No: 120
Ans: 1766/783

Page No: 13
Sum: +

S. No: 121
Ans: 719/585

Page No: 13
Sum: +

S. No: 122
Ans: 4367/2820

Page No: 13
Sum: +

S. No: 123
Ans: 1699/504

Page No: 13
Sum: +

S. No: 124
Ans: 6744/5605

Page No: 13
Sum: +

S. No: 125
Ans: 6019/4510

Page No: 13
Sum: +

S. No: 126
Ans: 283/110

Page No: 13
Sum: +

S. No: 127
Ans: 3763/1763

Page No: 13
Sum: +

S. No: 128
Ans: 51/26

Page No: 13
Sum: +

S. No: 129
Ans: 1027/261

Page No: 13
Sum: +

S. No: 130
Ans: 457/462

Page No: 14
Sum: +

S. No: 131
Ans: 7546/4575

Page No: 14
Sum: +

S. No: 132
Ans: 1105/276

Page No: 14
Sum: +

S. No: 133
Ans: 254/45

Page No: 14
Sum: +

S. No: 134
Ans: 493/144

Page No: 14
Sum: +

S. No: 135
Ans: 5363/3913

Page No: 14
Sum: +

S. No: 136
Ans: 91/30

Page No: 14
Sum: +

S. No: 137
Ans: 227/280

Page No: 14
Sum: +

S. No: 138
Ans: 2012/715

Page No: 14
Sum: +

S. No: 139
Ans: 67/87

Page No: 14
Sum: +

S. No: 140
Ans: 1886/297

Answers

Page No: 15
Sum: +

S. No: 141
Ans: 408/155

Page No: 15
Sum: +

S. No: 142
Ans: 4723/1650

Page No: 15
Sum: +

S. No: 143
Ans: 269/210

Page No: 15
Sum: +

S. No: 144
Ans: 1617/230

Page No: 15
Sum: +

S. No: 145
Ans: 239/60

Page No: 15
Sum: +

S. No: 146
Ans: 4418/1591

Page No: 15
Sum: +

S. No: 147
Ans: 594/247

Page No: 15
Sum: +

S. No: 148
Ans: 170/21

Page No: 15
Sum: +

S. No: 149
Ans: 130/51

Page No: 15
Sum: +

S. No: 150
Ans: 4736/6873

Page No: 16
Sum: +

S. No: 151
Ans: 3188/3901

Page No: 16
Sum: +

S. No: 152
Ans: 8381/5112

Page No: 16
Sum: +

S. No: 153
Ans: 3992/1443

Page No: 16
Sum: +

S. No: 154
Ans: 1685/702

Page No: 16
Sum: +

S. No: 155
Ans: 2777/3201

Page No: 16
Sum: +

S. No: 156
Ans: 534/155

Page No: 16
Sum: +

S. No: 157
Ans: 1714/869

Page No: 16
Sum: +

S. No: 158
Ans: 1549/660

Page No: 16
Sum: +

S. No: 159
Ans: 10482/4067

Page No: 16
Sum: +

S. No: 160
Ans: 1007/940

Page No: 17
Sum: +

S. No: 161
Ans: 1171/153

Page No: 17
Sum: +

S. No: 162
Ans: 213/140

Page No: 17
Sum: +

S. No: 163
Ans: 1427/255

Page No: 17
Sum: +

S. No: 164
Ans: 2675/1428

Page No: 17
Sum: +

S. No: 165
Ans: 512/215

Page No: 17
Sum: +

S. No: 166
Ans: 9619/2214

Page No: 17
Sum: +

S. No: 167
Ans: 4835/646

Page No: 17
Sum: +

S. No: 168
Ans: 1613/912

Answers

Page No: 17
Sum: +

S. No: 169
Ans: 9901/8184

Page No: 17
Sum: +

S. No: 170
Ans: 1801/528

Page No: 18
Sum: +

S. No: 171
Ans: 958/441

Page No: 18
Sum: +

S. No: 172
Ans: 275/57

Page No: 18
Sum: +

S. No: 173
Ans: 7547/1940

Page No: 18
Sum: +

S. No: 174
Ans: 731/530

Page No: 18
Sum: +

S. No: 175
Ans: 737/344

Page No: 18
Sum: +

S. No: 176
Ans: 2039/340

Page No: 18
Sum: +

S. No: 177
Ans: 1321/990

Page No: 18
Sum: +

S. No: 178
Ans: 6533/2074

Page No: 18
Sum: +

S. No: 179
Ans: 3086/1113

Page No: 18
Sum: +

S. No: 180
Ans: 503/295

Page No: 19
Sum: +

S. No: 181
Ans: 3467/893

Page No: 19
Sum: +

S. No: 182
Ans: 5330/931

Page No: 19
Sum: +

S. No: 183
Ans: 16472/7857

Page No: 19
Sum: +

S. No: 184
Ans: 325/567

Page No: 19
Sum: +

S. No: 185
Ans: 1555/264

Page No: 19
Sum: +

S. No: 186
Ans: 11149/6674

Page No: 19
Sum: +

S. No: 187
Ans: 4675/4662

Page No: 19
Sum: +

S. No: 188
Ans: 5763/1424

Page No: 19
Sum: +

S. No: 189
Ans: 1837/464

Page No: 19
Sum: +

S. No: 190
Ans: 8412/7031

Page No: 20
Sum: +

S. No: 191
Ans: 2632/1653

Page No: 20
Sum: +

S. No: 192
Ans: 1319/696

Page No: 20
Sum: +

S. No: 193
Ans: 113/98

Page No: 20
Sum: +

S. No: 194
Ans: 1200/737

Page No: 20
Sum: +

S. No: 195
Ans: 790/341

Page No: 20
Sum: +

S. No: 196
Ans: 136/45

Answers

Page No: 20
Sum: +

S. No: 197
Ans: 259/120

Page No: 20
Sum: +

S. No: 198
Ans: 514/357

Page No: 20
Sum: +

S. No: 199
Ans: 2724/1265

Page No: 20
Sum: +

S. No: 200
Ans: 2315/1479

Page No: 21
Sum: +

S. No: 201
Ans: 412/195

Page No: 21
Sum: +

S. No: 202
Ans: 2159/558

Page No: 21
Sum: +

S. No: 203
Ans: 149/11

Page No: 21
Sum: +

S. No: 204
Ans: 1295/1104

Page No: 21
Sum: +

S. No: 205
Ans: 970/891

Page No: 21
Sum: +

S. No: 206
Ans: 361/96

Page No: 21
Sum: +

S. No: 207
Ans: 1766/435

Page No: 21
Sum: +

S. No: 208
Ans: 694/195

Page No: 21
Sum: +

S. No: 209
Ans: 1776/1007

Page No: 21
Sum: +

S. No: 210
Ans: 3257/4005

Page No: 22
Sum: +

S. No: 211
Ans: 422/143

Page No: 22
Sum: +

S. No: 212
Ans: 821/429

Page No: 22
Sum: +

S. No: 213
Ans: 8053/2772

Page No: 22
Sum: +

S. No: 214
Ans: 8935/5044

Page No: 22
Sum: +

S. No: 215
Ans: 7487/4218

Page No: 22
Sum: +

S. No: 216
Ans: 2473/851

Page No: 22
Sum: +

S. No: 217
Ans: 1456/517

Page No: 22
Sum: +

S. No: 218
Ans: 716/485

Page No: 22
Sum: +

S. No: 219
Ans: 929/396

Page No: 22
Sum: +

S. No: 220
Ans: 5443/1105

Page No: 23
Sum: +

S. No: 221
Ans: 3292/1353

Page No: 23
Sum: +

S. No: 222
Ans: 1435/459

Page No: 23
Sum: +

S. No: 223
Ans: 497/186

Page No: 23
Sum: +

S. No: 224
Ans: 382/135

Answers

Page No: 23
Sum: +

S. No: 225
Ans: 766/649

Page No: 23
Sum: +

S. No: 226
Ans: 313/140

Page No: 23
Sum: +

S. No: 227
Ans: 218/39

Page No: 23
Sum: +

S. No: 228
Ans: 1109/570

Page No: 23
Sum: +

S. No: 229
Ans: 5162/2331

Page No: 23
Sum: +

S. No: 230
Ans: 4890/3311

Page No: 24
Sum: +

S. No: 231
Ans: 116/63

Page No: 24
Sum: +

S. No: 232
Ans: 4034/1591

Page No: 24
Sum: +

S. No: 233
Ans: 51/14

Page No: 24
Sum: +

S. No: 234
Ans: 6709/1479

Page No: 24
Sum: +

S. No: 235
Ans: 638/305

Page No: 24
Sum: +

S. No: 236
Ans: 79/184

Page No: 24
Sum: +

S. No: 237
Ans: 2393/2175

Page No: 24
Sum: +

S. No: 238
Ans: 2216/795

Page No: 24
Sum: +

S. No: 239
Ans: 1388/495

Page No: 24
Sum: +

S. No: 240
Ans: 1501/582

Page No: 25
Sum: +

S. No: 241
Ans: 2641/522

Page No: 25
Sum: +

S. No: 242
Ans: 1217/1015

Page No: 25
Sum: +

S. No: 243
Ans: 1119/430

Page No: 25
Sum: +

S. No: 244
Ans: 619/441

Page No: 25
Sum: +

S. No: 245
Ans: 2543/589

Page No: 25
Sum: +

S. No: 246
Ans: 1614/469

Page No: 25
Sum: +

S. No: 247
Ans: 243/88

Page No: 25
Sum: +

S. No: 248
Ans: 595/426

Page No: 25
Sum: +

S. No: 249
Ans: 4392/1421

Page No: 25
Sum: +

S. No: 250
Ans: 6958/2655

Page No: 26
Sum: +

S. No: 251
Ans: 9747/4118

Page No: 26
Sum: +

S. No: 252
Ans: 5358/3827

Answers

Page No: 26
Sum: +

S. No: 253
Ans: 1984/781

Page No: 26
Sum: +

S. No: 254
Ans: 261/38

Page No: 26
Sum: +

S. No: 255
Ans: 10160/2619

Page No: 26
Sum: +

S. No: 256
Ans: 995/931

Page No: 26
Sum: +

S. No: 257
Ans: 1628/779

Page No: 26
Sum: +

S. No: 258
Ans: 2399/1470

Page No: 26
Sum: +

S. No: 259
Ans: 6427/2440

Page No: 26
Sum: +

S. No: 260
Ans: 74/15

Page No: 27
Sum: +

S. No: 261
Ans: 217/40

Page No: 27
Sum: +

S. No: 262
Ans: 1319/189

Page No: 27
Sum: +

S. No: 263
Ans: 808/1053

Page No: 27
Sum: +

S. No: 264
Ans: 7069/2482

Page No: 27
Sum: +

S. No: 265
Ans: 677/340

Page No: 27
Sum: +

S. No: 266
Ans: 11771/7980

Page No: 27
Sum: +

S. No: 267
Ans: 10624/4389

Page No: 27
Sum: +

S. No: 268
Ans: 10813/4628

Page No: 27
Sum: +

S. No: 269
Ans: 2069/408

Page No: 27
Sum: +

S. No: 270
Ans: 3234/2795

Page No: 28
Sum: +

S. No: 271
Ans: 3545/3243

Page No: 28
Sum: +

S. No: 272
Ans: 226/195

Page No: 28
Sum: +

S. No: 273
Ans: 409/150

Page No: 28
Sum: +

S. No: 274
Ans: 3794/1275

Page No: 28
Sum: +

S. No: 275
Ans: 232/85

Page No: 28
Sum: +

S. No: 276
Ans: 5419/1340

Page No: 28
Sum: +

S. No: 277
Ans: 10354/2175

Page No: 28
Sum: +

S. No: 278
Ans: 4099/3850

Page No: 28
Sum: +

S. No: 279
Ans: 83/72

Page No: 28
Sum: +

S. No: 280
Ans: 2997/2581

Answers

Page No: 29
Sum: +

S. No: 281
Ans: 6517/1935

Page No: 29
Sum: +

S. No: 282
Ans: 5350/2769

Page No: 29
Sum: +

S. No: 283
Ans: 454/105

Page No: 29
Sum: +

S. No: 284
Ans: 323/75

Page No: 29
Sum: +

S. No: 285
Ans: 5101/3792

Page No: 29
Sum: +

S. No: 286
Ans: 13811/6150

Page No: 29
Sum: +

S. No: 287
Ans: 1517/868

Page No: 29
Sum: +

S. No: 288
Ans: 433/238

Page No: 29
Sum: +

S. No: 289
Ans: 281/140

Page No: 29
Sum: +

S. No: 290
Ans: 1997/1290

Page No: 30
Sum: +

S. No: 291
Ans: 356/165

Page No: 30
Sum: +

S. No: 292
Ans: 125/14

Page No: 30
Sum: +

S. No: 293
Ans: 481/196

Page No: 30
Sum: +

S. No: 294
Ans: 1131/352

Page No: 30
Sum: +

S. No: 295
Ans: 6004/3395

Page No: 30
Sum: +

S. No: 296
Ans: 1075/611

Page No: 30
Sum: +

S. No: 297
Ans: 1852/621

Page No: 30
Sum: +

S. No: 298
Ans: 733/124

Page No: 30
Sum: +

S. No: 299
Ans: 404/45

Page No: 30
Sum: +

S. No: 300
Ans: 547/180

Page No: 31
Sum: +

S. No: 301
Ans: 2147/984

Page No: 31
Sum: +

S. No: 302
Ans: 3481/852

Page No: 31
Sum: +

S. No: 303
Ans: 839/221

Page No: 31
Sum: +

S. No: 304
Ans: 8243/836

Page No: 31
Sum: +

S. No: 305
Ans: 4914/3403

Page No: 31
Sum: +

S. No: 306
Ans: 671/288

Page No: 31
Sum: +

S. No: 307
Ans: 340/77

Page No: 31
Sum: +

S. No: 308
Ans: 10393/3483

Answers

Page No: 31 Sum: + *S. No: 309* **Ans: 997/340**	Page No: 31 Sum: + *S. No: 310* **Ans: 10809/3827**	Page No: 32 Sum: + *S. No: 311* **Ans: 1023/602**	Page No: 32 Sum: + *S. No: 312* **Ans: 4428/899**
Page No: 32 Sum: + *S. No: 313* **Ans: 349/195**	Page No: 32 Sum: + *S. No: 314* **Ans: 4961/720**	Page No: 32 Sum: + *S. No: 315* **Ans: 1331/688**	Page No: 32 Sum: + *S. No: 316* **Ans: 85/77**
Page No: 32 Sum: + *S. No: 317* **Ans: 3891/782**	Page No: 32 Sum: + *S. No: 318* **Ans: 1140/143**	Page No: 32 Sum: + *S. No: 319* **Ans: 251/96**	Page No: 32 Sum: + *S. No: 320* **Ans: 6743/4346**
Page No: 33 Sum: + *S. No: 321* **Ans: 10669/4108**	Page No: 33 Sum: + *S. No: 322* **Ans: 3883/3444**	Page No: 33 Sum: + *S. No: 323* **Ans: 111/85**	Page No: 33 Sum: + *S. No: 324* **Ans: 3907/1225**
Page No: 33 Sum: + *S. No: 325* **Ans: 3634/1729**	Page No: 33 Sum: + *S. No: 326* **Ans: 3313/4872**	Page No: 33 Sum: + *S. No: 327* **Ans: 176/23**	Page No: 33 Sum: + *S. No: 328* **Ans: 3818/1995**
Page No: 33 Sum: + *S. No: 329* **Ans: 939/440**	Page No: 33 Sum: + *S. No: 330* **Ans: 4863/4462**	Page No: 34 Sum: + *S. No: 331* **Ans: 9077/3230**	Page No: 34 Sum: + *S. No: 332* **Ans: 647/376**
Page No: 34 Sum: + *S. No: 333* **Ans: 2998/1909**	Page No: 34 Sum: + *S. No: 334* **Ans: 5897/2418**	Page No: 34 Sum: + *S. No: 335* **Ans: 247/66**	Page No: 34 Sum: + *S. No: 336* **Ans: 3651/910**

Answers

Page No: 34
Sum: +

S. No: 337
Ans: 3629/2716

Page No: 34
Sum: +

S. No: 338
Ans: 5353/3283

Page No: 34
Sum: +

S. No: 339
Ans: 85/18

Page No: 34
Sum: +

S. No: 340
Ans: 2377/1504

Page No: 35
Sum: +

S. No: 341
Ans: 87/41

Page No: 35
Sum: +

S. No: 342
Ans: 8579/9506

Page No: 35
Sum: +

S. No: 343
Ans: 89/21

Page No: 35
Sum: +

S. No: 344
Ans: 8933/6072

Page No: 35
Sum: +

S. No: 345
Ans: 87/14

Page No: 35
Sum: +

S. No: 346
Ans: 93/50

Page No: 35
Sum: +

S. No: 347
Ans: 541/186

Page No: 35
Sum: +

S. No: 348
Ans: 568/85

Page No: 35
Sum: +

S. No: 349
Ans: 1441/156

Page No: 35
Sum: +

S. No: 350
Ans: 326/75

Page No: 36
Sum: -

S. No: 351
Ans: -4/9

Page No: 36
Sum: -

S. No: 352
Ans: -1/5

Page No: 36
Sum: -

S. No: 353
Ans: 0

Page No: 36
Sum: -

S. No: 354
Ans: 7/9

Page No: 36
Sum: -

S. No: 355
Ans: 3/7

Page No: 36
Sum: -

S. No: 356
Ans: -5/3

Page No: 36
Sum: -

S. No: 357
Ans: 1/9

Page No: 36
Sum: -

S. No: 358
Ans: 1/5

Page No: 36
Sum: -

S. No: 359
Ans: 0

Page No: 36
Sum: -

S. No: 360
Ans: 5/2

Page No: 37
Sum: -

S. No: 361
Ans: 2

Page No: 37
Sum: -

S. No: 362
Ans: 3/2

Page No: 37
Sum: -

S. No: 363
Ans: 5/2

Page No: 37
Sum: -

S. No: 364
Ans: 1

Answers

Page No: 37
Sum: -

S. No: 365
Ans: 2/7

Page No: 37
Sum: -

S. No: 366
Ans: 1/9

Page No: 37
Sum: -

S. No: 367
Ans: 0

Page No: 37
Sum: -

S. No: 368
Ans: -4/7

Page No: 37
Sum: -

S. No: 369
Ans: 0

Page No: 37
Sum: -

S. No: 370
Ans: -1

Page No: 38
Sum: -

S. No: 371
Ans: 44/43

Page No: 38
Sum: -

S. No: 372
Ans: -1/38

Page No: 38
Sum: -

S. No: 373
Ans: -21/17

Page No: 38
Sum: -

S. No: 374
Ans: -13/66

Page No: 38
Sum: -

S. No: 375
Ans: 10/9

Page No: 38
Sum: -

S. No: 376
Ans: -29/83

Page No: 38
Sum: -

S. No: 377
Ans: 31/48

Page No: 38
Sum: -

S. No: 378
Ans: -11/6

Page No: 38
Sum: -

S. No: 379
Ans: -1/47

Page No: 38
Sum: -

S. No: 380
Ans: -5/39

Page No: 39
Sum: -

S. No: 381
Ans: 7/50

Page No: 39
Sum: -

S. No: 382
Ans: 17/18

Page No: 39
Sum: -

S. No: 383
Ans: 47/15

Page No: 39
Sum: -

S. No: 384
Ans: -15/71

Page No: 39
Sum: -

S. No: 385
Ans: 37/71

Page No: 39
Sum: -

S. No: 386
Ans: -3/7

Page No: 39
Sum: -

S. No: 387
Ans: 8/11

Page No: 39
Sum: -

S. No: 388
Ans: 11/10

Page No: 39
Sum: -

S. No: 389
Ans: 77/69

Page No: 39
Sum: -

S. No: 390
Ans: -25/14

Page No: 40
Sum: -

S. No: 391
Ans: 27/91

Page No: 40
Sum: -

S. No: 392
Ans: 31/10

Answers

Page No: 40
Sum: -

S. No: 393
Ans: 7/95

Page No: 40
Sum: -

S. No: 394
Ans: 3/8

Page No: 40
Sum: -

S. No: 395
Ans: -20/11

Page No: 40
Sum: -

S. No: 396
Ans: -59/79

Page No: 40
Sum: -

S. No: 397
Ans: 25/96

Page No: 40
Sum: -

S. No: 398
Ans: -29/75

Page No: 40
Sum: -

S. No: 399
Ans: -63/73

Page No: 40
Sum: -

S. No: 400
Ans: 4/15

Page No: 41
Sum: -

S. No: 401
Ans: -33/97

Page No: 41
Sum: -

S. No: 402
Ans: -31/27

Page No: 41
Sum: -

S. No: 403
Ans: -1/19

Page No: 41
Sum: -

S. No: 404
Ans: -6/35

Page No: 41
Sum: -

S. No: 405
Ans: 21/64

Page No: 41
Sum: -

S. No: 406
Ans: 35/62

Page No: 41
Sum: -

S. No: 407
Ans: 59/30

Page No: 41
Sum: -

S. No: 408
Ans: 41/62

Page No: 41
Sum: -

S. No: 409
Ans: 1/9

Page No: 41
Sum: -

S. No: 410
Ans: 51/91

Page No: 42
Sum: -

S. No: 411
Ans: -1/15

Page No: 42
Sum: -

S. No: 412
Ans: -16/15

Page No: 42
Sum: -

S. No: 413
Ans: 3/14

Page No: 42
Sum: -

S. No: 414
Ans: -11/19

Page No: 42
Sum: -

S. No: 415
Ans: 9/10

Page No: 42
Sum: -

S. No: 416
Ans: 28/17

Page No: 42
Sum: -

S. No: 417
Ans: -33/40

Page No: 42
Sum: -

S. No: 418
Ans: 6/13

Page No: 42
Sum: -

S. No: 419
Ans: -23/63

Page No: 42
Sum: -

S. No: 420
Ans: -11/20

Answers

Page No: 43
Sum: -

S. No: 421
Ans: -36/71

Page No: 43
Sum: -

S. No: 422
Ans: -45/34

Page No: 43
Sum: -

S. No: 423
Ans: 2/87

Page No: 43
Sum: -

S. No: 424
Ans: -7/81

Page No: 43
Sum: -

S. No: 425
Ans: 7/55

Page No: 43
Sum: -

S. No: 426
Ans: -22/27

Page No: 43
Sum: -

S. No: 427
Ans: -27/94

Page No: 43
Sum: -

S. No: 428
Ans: -1/2

Page No: 43
Sum: -

S. No: 429
Ans: -3/13

Page No: 43
Sum: -

S. No: 430
Ans: 87/26

Page No: 44
Sum: -

S. No: 431
Ans: 12/5

Page No: 44
Sum: -

S. No: 432
Ans: -15/4

Page No: 44
Sum: -

S. No: 433
Ans: 11/24

Page No: 44
Sum: -

S. No: 434
Ans: 35/4

Page No: 44
Sum: -

S. No: 435
Ans: -19/10

Page No: 44
Sum: -

S. No: 436
Ans: 55/56

Page No: 44
Sum: -

S. No: 437
Ans: 10/9

Page No: 44
Sum: -

S. No: 438
Ans: -11/12

Page No: 44
Sum: -

S. No: 439
Ans: 0

Page No: 44
Sum: -

S. No: 440
Ans: -1

Page No: 45
Sum: -

S. No: 441
Ans: 1/12

Page No: 45
Sum: -

S. No: 442
Ans: 11/6

Page No: 45
Sum: -

S. No: 443
Ans: -25/7

Page No: 45
Sum: -

S. No: 444
Ans: -7/8

Page No: 45
Sum: -

S. No: 445
Ans: 4/15

Page No: 45
Sum: -

S. No: 446
Ans: -17/56

Page No: 45
Sum: -

S. No: 447
Ans: 15/8

Page No: 45
Sum: -

S. No: 448
Ans: 8/7

Answers

Page No: 45 Sum: - *S. No: 449* **Ans: 23/5**	Page No: 45 Sum: - *S. No: 450* **Ans: 1/6**	Page No: 46 Sum: - *S. No: 451* **Ans: -3164/3237**	Page No: 46 Sum: - *S. No: 452* **Ans: -2269/4070**
Page No: 46 Sum: - *S. No: 453* **Ans: -11/30**	Page No: 46 Sum: - *S. No: 454* **Ans: 359/231**	Page No: 46 Sum: - *S. No: 455* **Ans: 35/68**	Page No: 46 Sum: - *S. No: 456* **Ans: 2177/3960**
Page No: 46 Sum: - *S. No: 457* **Ans: -963/602**	Page No: 46 Sum: - *S. No: 458* **Ans: -23/180**	Page No: 46 Sum: - *S. No: 459* **Ans: 19/18**	Page No: 46 Sum: - *S. No: 460* **Ans: -343/682**
Page No: 47 Sum: - *S. No: 461* **Ans: -871/220**	Page No: 47 Sum: - *S. No: 462* **Ans: 1/7**	Page No: 47 Sum: - *S. No: 463* **Ans: -117/77**	Page No: 47 Sum: - *S. No: 464* **Ans: -1933/1887**
Page No: 47 Sum: - *S. No: 465* **Ans: -323/154**	Page No: 47 Sum: - *S. No: 466* **Ans: -398/187**	Page No: 47 Sum: - *S. No: 467* **Ans: 167/60**	Page No: 47 Sum: - *S. No: 468* **Ans: 37/90**
Page No: 47 Sum: - *S. No: 469* **Ans: 393/65**	Page No: 47 Sum: - *S. No: 470* **Ans: 119/783**	Page No: 48 Sum: - *S. No: 471* **Ans: -464/585**	Page No: 48 Sum: - *S. No: 472* **Ans: 1367/2820**
Page No: 48 Sum: - *S. No: 473* **Ans: 275/504**	Page No: 48 Sum: - *S. No: 474* **Ans: 1044/5605**	Page No: 48 Sum: - *S. No: 475* **Ans: 4149/4510**	Page No: 48 Sum: - *S. No: 476* **Ans: -29/55**

Answers

Page No: 48
Sum: -

S. No: 477
Ans: 1139/1763

Page No: 48
Sum: -

S. No: 478
Ans: 25/26

Page No: 48
Sum: -

S. No: 479
Ans: -737/261

Page No: 48
Sum: -

S. No: 480
Ans: 107/462

Page No: 49
Sum: -

S. No: 481
Ans: 104/4575

Page No: 49
Sum: -

S. No: 482
Ans: -781/276

Page No: 49
Sum: -

S. No: 483
Ans: -196/45

Page No: 49
Sum: -

S. No: 484
Ans: 29/144

Page No: 49
Sum: -

S. No: 485
Ans: -461/3913

Page No: 49
Sum: -

S. No: 486
Ans: 11/30

Page No: 49
Sum: -

S. No: 487
Ans: 43/280

Page No: 49
Sum: -

S. No: 488
Ans: 738/715

Page No: 49
Sum: -

S. No: 489
Ans: -11/29

Page No: 49
Sum: -

S. No: 490
Ans: 1732/297

Page No: 50
Sum: -

S. No: 491
Ans: -88/155

Page No: 50
Sum: -

S. No: 492
Ans: -2677/1650

Page No: 50
Sum: -

S. No: 493
Ans: 137/210

Page No: 50
Sum: -

S. No: 494
Ans: -283/230

Page No: 50
Sum: -

S. No: 495
Ans: -161/60

Page No: 50
Sum: -

S. No: 496
Ans: 2612/1591

Page No: 50
Sum: -

S. No: 497
Ans: -290/247

Page No: 50
Sum: -

S. No: 498
Ans: 118/21

Page No: 50
Sum: -

S. No: 499
Ans: -8/51

Page No: 50
Sum: -

S. No: 500
Ans: -2050/6873

Page No: 51
Sum: -

S. No: 501
Ans: 2358/3901

Page No: 51
Sum: -

S. No: 502
Ans: 2275/5112

Page No: 51
Sum: -

S. No: 503
Ans: 1624/1443

Page No: 51
Sum: -

S. No: 504
Ans: -151/702

Answers

Page No: 51 Sum: - *S. No: 505* **Ans: -61/3201**	Page No: 51 Sum: - *S. No: 506* **Ans: -21/155**	Page No: 51 Sum: - *S. No: 507* **Ans: -134/869**	Page No: 51 Sum: - *S. No: 508* **Ans: -79/660**
Page No: 51 Sum: - *S. No: 509* **Ans: -5288/4067**	Page No: 51 Sum: - *S. No: 510* **Ans: 27/940**	Page No: 52 Sum: - *S. No: 511* **Ans: -478/153**	Page No: 52 Sum: - *S. No: 512* **Ans: 157/140**
Page No: 52 Sum: - *S. No: 513* **Ans: -1259/255**	Page No: 52 Sum: - *S. No: 514* **Ans: 601/1428**	Page No: 52 Sum: - *S. No: 515* **Ans: -47/215**	Page No: 52 Sum: - *S. No: 516* **Ans: -4813/2214**
Page No: 52 Sum: - *S. No: 517* **Ans: 2081/646**	Page No: 52 Sum: - *S. No: 518* **Ans: -781/912**	Page No: 52 Sum: - *S. No: 519* **Ans: -8141/8184**	Page No: 52 Sum: - *S. No: 520* **Ans: -1559/528**
Page No: 53 Sum: - *S. No: 521* **Ans: 22/441**	Page No: 53 Sum: - *S. No: 522* **Ans: 199/57**	Page No: 53 Sum: - *S. No: 523* **Ans: 6227/1940**	Page No: 53 Sum: - *S. No: 524* **Ans: -11/530**
Page No: 53 Sum: - *S. No: 525* **Ans: -479/344**	Page No: 53 Sum: - *S. No: 526* **Ans: 1429/340**	Page No: 53 Sum: - *S. No: 527* **Ans: -199/990**	Page No: 53 Sum: - *S. No: 528* **Ans: 2861/2074**
Page No: 53 Sum: - *S. No: 529* **Ans: -1072/1113**	Page No: 53 Sum: - *S. No: 530* **Ans: 267/295**	Page No: 54 Sum: - *S. No: 531* **Ans: 1985/893**	Page No: 54 Sum: - *S. No: 532* **Ans: 3392/931**

Answers

Page No: 54
Sum: -

S. No: 533
Ans: 2540/7857

Page No: 54
Sum: -

S. No: 534
Ans: -109/567

Page No: 54
Sum: -

S. No: 535
Ans: -1181/264

Page No: 54
Sum: -

S. No: 536
Ans: -1635/6674

Page No: 54
Sum: -

S. No: 537
Ans: 2281/4662

Page No: 54
Sum: -

S. No: 538
Ans: 4739/1424

Page No: 54
Sum: -

S. No: 539
Ans: 851/464

Page No: 54
Sum: -

S. No: 540
Ans: -3514/7031

Page No: 55
Sum: -

S. No: 541
Ans: -370/1653

Page No: 55
Sum: -

S. No: 542
Ans: -237/232

Page No: 55
Sum: -

S. No: 543
Ans: -85/98

Page No: 55
Sum: -

S. No: 544
Ans: 408/737

Page No: 55
Sum: -

S. No: 545
Ans: 46/341

Page No: 55
Sum: -

S. No: 546
Ans: 28/45

Page No: 55
Sum: -

S. No: 547
Ans: 29/120

Page No: 55
Sum: -

S. No: 548
Ans: -4/357

Page No: 55
Sum: -

S. No: 549
Ans: -634/1265

Page No: 55
Sum: -

S. No: 550
Ans: -1561/1479

Page No: 56
Sum: -

S. No: 551
Ans: 316/195

Page No: 56
Sum: -

S. No: 552
Ans: 755/558

Page No: 56
Sum: -

S. No: 553
Ans: 5/11

Page No: 56
Sum: -

S. No: 554
Ans: -407/1104

Page No: 56
Sum: -

S. No: 555
Ans: 592/891

Page No: 56
Sum: -

S. No: 556
Ans: 113/96

Page No: 56
Sum: -

S. No: 557
Ans: -1076/435

Page No: 56
Sum: -

S. No: 558
Ans: 256/195

Page No: 56
Sum: -

S. No: 559
Ans: 504/1007

Page No: 56
Sum: -

S. No: 560
Ans: -1727/4005

Answers

Page No: 57
Sum: -

S. No: 561
Ans: -202/143

Page No: 57
Sum: -

S. No: 562
Ans: -128/429

Page No: 57
Sum: -

S. No: 563
Ans: -3805/2772

Page No: 57
Sum: -

S. No: 564
Ans: 7167/5044

Page No: 57
Sum: -

S. No: 565
Ans: -3343/4218

Page No: 57
Sum: -

S. No: 566
Ans: 1437/851

Page No: 57
Sum: -

S. No: 567
Ans: 565/517

Page No: 57
Sum: -

S. No: 568
Ans: -134/485

Page No: 57
Sum: -

S. No: 569
Ans: 83/396

Page No: 57
Sum: -

S. No: 570
Ans: -3267/1105

Page No: 58
Sum: -

S. No: 571
Ans: 668/1353

Page No: 58
Sum: -

S. No: 572
Ans: -347/459

Page No: 58
Sum: -

S. No: 573
Ans: 187/186

Page No: 58
Sum: -

S. No: 574
Ans: -212/135

Page No: 58
Sum: -

S. No: 575
Ans: 414/649

Page No: 58
Sum: -

S. No: 576
Ans: 23/140

Page No: 58
Sum: -

S. No: 577
Ans: -66/13

Page No: 58
Sum: -

S. No: 578
Ans: 449/570

Page No: 58
Sum: -

S. No: 579
Ans: 130/2331

Page No: 58
Sum: -

S. No: 580
Ans: 270/3311

Page No: 59
Sum: -

S. No: 581
Ans: 17/63

Page No: 59
Sum: -

S. No: 582
Ans: -1814/1591

Page No: 59
Sum: -

S. No: 583
Ans: 83/42

Page No: 59
Sum: -

S. No: 584
Ans: -3287/1479

Page No: 59
Sum: -

S. No: 585
Ans: 272/305

Page No: 59
Sum: -

S. No: 586
Ans: 33/184

Page No: 59
Sum: -

S. No: 587
Ans: -469/725

Page No: 59
Sum: -

S. No: 588
Ans: 116/795

Answers

Page No: 59
Sum: -

S. No: 589
Ans: -256/165

Page No: 59
Sum: -

S. No: 590
Ans: 1021/582

Page No: 60
Sum: -

S. No: 591
Ans: -2173/522

Page No: 60
Sum: -

S. No: 592
Ans: 463/1015

Page No: 60
Sum: -

S. No: 593
Ans: 1/430

Page No: 60
Sum: -

S. No: 594
Ans: -227/441

Page No: 60
Sum: -

S. No: 595
Ans: -681/589

Page No: 60
Sum: -

S. No: 596
Ans: -1334/469

Page No: 60
Sum: -

S. No: 597
Ans: 45/88

Page No: 60
Sum: -

S. No: 598
Ans: -115/426

Page No: 60
Sum: -

S. No: 599
Ans: -1060/1421

Page No: 60
Sum: -

S. No: 600
Ans: 1412/2655

Page No: 61
Sum: -

S. No: 601
Ans: 2323/4118

Page No: 61
Sum: -

S. No: 602
Ans: -1264/3827

Page No: 61
Sum: -

S. No: 603
Ans: -4/781

Page No: 61
Sum: -

S. No: 604
Ans: -111/38

Page No: 61
Sum: -

S. No: 605
Ans: 8270/2619

Page No: 61
Sum: -

S. No: 606
Ans: -83/931

Page No: 61
Sum: -

S. No: 607
Ans: -234/779

Page No: 61
Sum: -

S. No: 608
Ans: 1619/1470

Page No: 61
Sum: -

S. No: 609
Ans: -693/2440

Page No: 61
Sum: -

S. No: 610
Ans: -16/15

Page No: 62
Sum: -

S. No: 611
Ans: -127/40

Page No: 62
Sum: -

S. No: 612
Ans: 787/189

Page No: 62
Sum: -

S. No: 613
Ans: 2/1053

Page No: 62
Sum: -

S. No: 614
Ans: 4757/2482

Page No: 62
Sum: -

S. No: 615
Ans: -547/340

Page No: 62
Sum: -

S. No: 616
Ans: 4379/7980

Answers

Page No: 62
Sum: -

S. No: 617
Ans: 4468/4389

Page No: 62
Sum: -

S. No: 618
Ans: 2181/4628

Page No: 62
Sum: -

S. No: 619
Ans: -413/408

Page No: 62
Sum: -

S. No: 620
Ans: -894/2795

Page No: 63
Sum: -

S. No: 621
Ans: 1061/3243

Page No: 63
Sum: -

S. No: 622
Ans: 44/195

Page No: 63
Sum: -

S. No: 623
Ans: -113/150

Page No: 63
Sum: -

S. No: 624
Ans: -694/1275

Page No: 63
Sum: -

S. No: 625
Ans: 28/85

Page No: 63
Sum: -

S. No: 626
Ans: 4899/1340

Page No: 63
Sum: -

S. No: 627
Ans: 5654/2175

Page No: 63
Sum: -

S. No: 628
Ans: 1599/3850

Page No: 63
Sum: -

S. No: 629
Ans: 257/792

Page No: 63
Sum: -

S. No: 630
Ans: 2107/2581

Page No: 64
Sum: -

S. No: 631
Ans: -2303/1935

Page No: 64
Sum: -

S. No: 632
Ans: 46/2769

Page No: 64
Sum: -

S. No: 633
Ans: 104/105

Page No: 64
Sum: -

S. No: 634
Ans: 127/75

Page No: 64
Sum: -

S. No: 635
Ans: 2099/3792

Page No: 64
Sum: -

S. No: 636
Ans: 2261/6150

Page No: 64
Sum: -

S. No: 637
Ans: 901/868

Page No: 64
Sum: -

S. No: 638
Ans: 111/238

Page No: 64
Sum: -

S. No: 639
Ans: -17/420

Page No: 64
Sum: -

S. No: 640
Ans: -497/1290

Page No: 65
Sum: -

S. No: 641
Ans: -4/165

Page No: 65
Sum: -

S. No: 642
Ans: 453/70

Page No: 65
Sum: -

S. No: 643
Ans: -9/196

Page No: 65
Sum: -

S. No: 644
Ans: 53/352

Answers

Page No: 65
Sum: -

S. No: 645
Ans: -1174/3395

Page No: 65
Sum: -

S. No: 646
Ans: 711/611

Page No: 65
Sum: -

S. No: 647
Ans: -862/621

Page No: 65
Sum: -

S. No: 648
Ans: 693/124

Page No: 65
Sum: -

S. No: 649
Ans: 199/45

Page No: 65
Sum: -

S. No: 650
Ans: -227/180

Page No: 66
Sum: -

S. No: 651
Ans: 589/984

Page No: 66
Sum: -

S. No: 652
Ans: -3193/852

Page No: 66
Sum: -

S. No: 653
Ans: 669/221

Page No: 66
Sum: -

S. No: 654
Ans: -6197/836

Page No: 66
Sum: -

S. No: 655
Ans: -896/3403

Page No: 66
Sum: -

S. No: 656
Ans: -31/288

Page No: 66
Sum: -

S. No: 657
Ans: 144/77

Page No: 66
Sum: -

S. No: 658
Ans: 2567/3483

Page No: 66
Sum: -

S. No: 659
Ans: 227/340

Page No: 66
Sum: -

S. No: 660
Ans: 2897/3827

Page No: 67
Sum: -

S. No: 661
Ans: -335/602

Page No: 67
Sum: -

S. No: 662
Ans: -1586/899

Page No: 67
Sum: -

S. No: 663
Ans: -217/390

Page No: 67
Sum: -

S. No: 664
Ans: -3409/720

Page No: 67
Sum: -

S. No: 665
Ans: -131/688

Page No: 67
Sum: -

S. No: 666
Ans: -41/77

Page No: 67
Sum: -

S. No: 667
Ans: 525/782

Page No: 67
Sum: -

S. No: 668
Ans: 906/143

Page No: 67
Sum: -

S. No: 669
Ans: 53/96

Page No: 67
Sum: -

S. No: 670
Ans: 2479/4346

Page No: 68
Sum: -

S. No: 671
Ans: -3077/4108

Page No: 68
Sum: -

S. No: 672
Ans: -1447/3444

Answers

Page No: 68
Sum: -

S. No: 673
Ans: 59/85

Page No: 68
Sum: -

S. No: 674
Ans: 307/1225

Page No: 68
Sum: -

S. No: 675
Ans: -2190/1729

Page No: 68
Sum: -

S. No: 676
Ans: -2081/4872

Page No: 68
Sum: -

S. No: 677
Ans: -315/46

Page No: 68
Sum: -

S. No: 678
Ans: -2908/1995

Page No: 68
Sum: -

S. No: 679
Ans: 741/440

Page No: 68
Sum: -

S. No: 680
Ans: 13/4462

Page No: 69
Sum: -

S. No: 681
Ans: -4793/3230

Page No: 69
Sum: -

S. No: 682
Ans: 199/376

Page No: 69
Sum: -

S. No: 683
Ans: -176/1909

Page No: 69
Sum: -

S. No: 684
Ans: 2101/2418

Page No: 69
Sum: -

S. No: 685
Ans: -37/66

Page No: 69
Sum: -

S. No: 686
Ans: 1991/910

Page No: 69
Sum: -

S. No: 687
Ans: 1221/2716

Page No: 69
Sum: -

S. No: 688
Ans: -3075/3283

Page No: 69
Sum: -

S. No: 689
Ans: -59/18

Page No: 69
Sum: -

S. No: 690
Ans: 1289/1504

Page No: 70
Sum: -

S. No: 691
Ans: 5/41

Page No: 70
Sum: -

S. No: 692
Ans: -151/9506

Page No: 70
Sum: -

S. No: 693
Ans: 47/21

Page No: 70
Sum: -

S. No: 694
Ans: 2245/6072

Page No: 70
Sum: -

S. No: 695
Ans: -73/14

Page No: 70
Sum: -

S. No: 696
Ans: -43/50

Page No: 70
Sum: -

S. No: 697
Ans: 479/186

Page No: 70
Sum: -

S. No: 698
Ans: -452/85

Page No: 70
Sum: -

S. No: 699
Ans: 791/156

Page No: 70
Sum: -

S. No: 700
Ans: 224/75

Answers

Page No: 71 Sum: x *S. No: 701* **Ans: 5/9**	Page No: 71 Sum: x *S. No: 702* **Ans: 42/25**	Page No: 71 Sum: x *S. No: 703* **Ans: 16/9**	Page No: 71 Sum: x *S. No: 704* **Ans: 2/9**
Page No: 71 Sum: x *S. No: 705* **Ans: 4/7**	Page No: 71 Sum: x *S. No: 706* **Ans: 4**	Page No: 71 Sum: x *S. No: 707* **Ans: 4/27**	Page No: 71 Sum: x *S. No: 708* **Ans: 4/5**
Page No: 71 Sum: x *S. No: 709* **Ans: 4**	Page No: 71 Sum: x *S. No: 710* **Ans: 7/2**	Page No: 72 Sum: x *S. No: 711* **Ans: 7/9**	Page No: 72 Sum: x *S. No: 712* **Ans: 1**
Page No: 72 Sum: x *S. No: 713* **Ans: 6**	Page No: 72 Sum: x *S. No: 714* **Ans: 30**	Page No: 72 Sum: x *S. No: 715* **Ans: 5/7**	Page No: 72 Sum: x *S. No: 716* **Ans: 56/81**
Page No: 72 Sum: x *S. No: 717* **Ans: 1/4**	Page No: 72 Sum: x *S. No: 718* **Ans: 3/7**	Page No: 72 Sum: x *S. No: 719* **Ans: 4/25**	Page No: 72 Sum: x *S. No: 720* **Ans: 12**
Page No: 73 Sum: x *S. No: 721* **Ans: 84/43**	Page No: 73 Sum: x *S. No: 722* **Ans: 1711/722**	Page No: 73 Sum: x *S. No: 723* **Ans: 5896/289**	Page No: 73 Sum: x *S. No: 724* **Ans: 664/363**
Page No: 73 Sum: x *S. No: 725* **Ans: 8/27**	Page No: 73 Sum: x *S. No: 726* **Ans: 2142/6889**	Page No: 73 Sum: x *S. No: 727* **Ans: 95/64**	Page No: 73 Sum: x *S. No: 728* **Ans: 445/64**

Answers

Page No: 73
Sum: x

S. No: 729
Ans: 600/2209

Page No: 73
Sum: x

S. No: 730
Ans: 3776/1521

Page No: 74
Sum: x

S. No: 731
Ans: 697/1250

Page No: 74
Sum: x

S. No: 732
Ans: 2303/1458

Page No: 74
Sum: x

S. No: 733
Ans: 26/5

Page No: 74
Sum: x

S. No: 734
Ans: 4104/5041

Page No: 74
Sum: x

S. No: 735
Ans: 5358/5041

Page No: 74
Sum: x

S. No: 736
Ans: 3337/3136

Page No: 74
Sum: x

S. No: 737
Ans: 6305/1936

Page No: 74
Sum: x

S. No: 738
Ans: 57/10

Page No: 74
Sum: x

S. No: 739
Ans: 20/69

Page No: 74
Sum: x

S. No: 740
Ans: 429/14

Page No: 75
Sum: x

S. No: 741
Ans: 6298/8281

Page No: 75
Sum: x

S. No: 742
Ans: 196/5

Page No: 75
Sum: x

S. No: 743
Ans: 1794/9025

Page No: 75
Sum: x

S. No: 744
Ans: 1813/1024

Page No: 75
Sum: x

S. No: 745
Ans: 679/11

Page No: 75
Sum: x

S. No: 746
Ans: 1886/6241

Page No: 75
Sum: x

S. No: 747
Ans: 13/48

Page No: 75
Sum: x

S. No: 748
Ans: 424/1875

Page No: 75
Sum: x

S. No: 749
Ans: 1558/5329

Page No: 75
Sum: x

S. No: 750
Ans: 539/1200

Page No: 76
Sum: x

S. No: 751
Ans: 4698/9409

Page No: 76
Sum: x

S. No: 752
Ans: 572/729

Page No: 76
Sum: x

S. No: 753
Ans: 837/5776

Page No: 76
Sum: x

S. No: 754
Ans: 432/1225

Page No: 76
Sum: x

S. No: 755
Ans: 207/256

Page No: 76
Sum: x

S. No: 756
Ans: 33/62

Answers

Page No: 76
Sum: x

S. No: 757
Ans: 527/150

Page No: 76
Sum: x

S. No: 758
Ans: 903/961

Page No: 76
Sum: x

S. No: 759
Ans: 664/1089

Page No: 76
Sum: x

S. No: 760
Ans: 4042/8281

Page No: 77
Sum: x

S. No: 761
Ans: 754/2025

Page No: 77
Sum: x

S. No: 762
Ans: 923/300

Page No: 77
Sum: x

S. No: 763
Ans: 115/448

Page No: 77
Sum: x

S. No: 764
Ans: 310/171

Page No: 77
Sum: x

S. No: 765
Ans: 1909/25

Page No: 77
Sum: x

S. No: 766
Ans: 124/17

Page No: 77
Sum: x

S. No: 767
Ans: 2349/800

Page No: 77
Sum: x

S. No: 768
Ans: 415/117

Page No: 77
Sum: x

S. No: 769
Ans: 290/567

Page No: 77
Sum: x

S. No: 770
Ans: 931/200

Page No: 78
Sum: x

S. No: 771
Ans: 2176/5041

Page No: 78
Sum: x

S. No: 772
Ans: 979/289

Page No: 78
Sum: x

S. No: 773
Ans: 215/841

Page No: 78
Sum: x

S. No: 774
Ans: 94/243

Page No: 78
Sum: x

S. No: 775
Ans: 376/605

Page No: 78
Sum: x

S. No: 776
Ans: 104/729

Page No: 78
Sum: x

S. No: 777
Ans: 1505/4418

Page No: 78
Sum: x

S. No: 778
Ans: 741/25

Page No: 78
Sum: x

S. No: 779
Ans: 5110/169

Page No: 78
Sum: x

S. No: 780
Ans: 297/169

Page No: 79
Sum: x

S. No: 781
Ans: 32/5

Page No: 79
Sum: x

S. No: 782
Ans: 27/2

Page No: 79
Sum: x

S. No: 783
Ans: 3/4

Page No: 79
Sum: x

S. No: 784
Ans: 9/4

Answers

Page No: 79
Sum: x

S. No: 785
Ans: 3/2

Page No: 79
Sum: x

S. No: 786
Ans: 9/56

Page No: 79
Sum: x

S. No: 787
Ans: 16/9

Page No: 79
Sum: x

S. No: 788
Ans: 35/24

Page No: 79
Sum: x

S. No: 789
Ans: 1

Page No: 79
Sum: x

S. No: 790
Ans: 2

Page No: 80
Sum: x

S. No: 791
Ans: 1/2

Page No: 80
Sum: x

S. No: 792
Ans: 5/3

Page No: 80
Sum: x

S. No: 793
Ans: 12/7

Page No: 80
Sum: x

S. No: 794
Ans: 1/8

Page No: 80
Sum: x

S. No: 795
Ans: 32/15

Page No: 80
Sum: x

S. No: 796
Ans: 3/56

Page No: 80
Sum: x

S. No: 797
Ans: 1/4

Page No: 80
Sum: x

S. No: 798
Ans: 12/7

Page No: 80
Sum: x

S. No: 799
Ans: 42/5

Page No: 80
Sum: x

S. No: 800
Ans: 2

Page No: 81
Sum: x

S. No: 801
Ans: 2291/1079

Page No: 81
Sum: x

S. No: 802
Ans: 2318/2035

Page No: 81
Sum: x

S. No: 803
Ans: 29/50

Page No: 81
Sum: x

S. No: 804
Ans: 712/231

Page No: 81
Sum: x

S. No: 805
Ans: 1425/1156

Page No: 81
Sum: x

S. No: 806
Ans: 3139/1980

Page No: 81
Sum: x

S. No: 807
Ans: 1989/602

Page No: 81
Sum: x

S. No: 808
Ans: 529/360

Page No: 81
Sum: x

S. No: 809
Ans: 116/27

Page No: 81
Sum: x

S. No: 810
Ans: 472/341

Page No: 82
Sum: x

S. No: 811
Ans: 129/88

Page No: 82
Sum: x

S. No: 812
Ans: 30/7

Answers

Page No: 82
Sum: x

S. No: 813
Ans: 170/77

Page No: 82
Sum: x

S. No: 814
Ans: 1128/629

Page No: 82
Sum: x

S. No: 815
Ans: 195/77

Page No: 82
Sum: x

S. No: 816
Ans: 329/187

Page No: 82
Sum: x

S. No: 817
Ans: 1189/120

Page No: 82
Sum: x

S. No: 818
Ans: 49/30

Page No: 82
Sum: x

S. No: 819
Ans: 62/65

Page No: 82
Sum: x

S. No: 820
Ans: 3965/3132

Page No: 83
Sum: x

S. No: 821
Ans: 119/540

Page No: 83
Sum: x

S. No: 822
Ans: 305/564

Page No: 83
Sum: x

S. No: 823
Ans: 4183/1512

Page No: 83
Sum: x

S. No: 824
Ans: 396/1121

Page No: 83
Sum: x

S. No: 825
Ans: 527/2255

Page No: 83
Sum: x

S. No: 826
Ans: 279/176

Page No: 83
Sum: x

S. No: 827
Ans: 1824/1763

Page No: 83
Sum: x

S. No: 828
Ans: 19/26

Page No: 83
Sum: x

S. No: 829
Ans: 490/261

Page No: 83
Sum: x

S. No: 830
Ans: 1175/5082

Page No: 84
Sum: x

S. No: 831
Ans: 17/25

Page No: 84
Sum: x

S. No: 832
Ans: 369/184

Page No: 84
Sum: x

S. No: 833
Ans: 29/9

Page No: 84
Sum: x

S. No: 834
Ans: 841/288

Page No: 84
Sum: x

S. No: 835
Ans: 1824/3913

Page No: 84
Sum: x

S. No: 836
Ans: 34/15

Page No: 84
Sum: x

S. No: 837
Ans: 621/3920

Page No: 84
Sum: x

S. No: 838
Ans: 245/143

Page No: 84
Sum: x

S. No: 839
Ans: 850/7569

Page No: 84
Sum: x

S. No: 840
Ans: 469/297

Answers

Page No: 85
Sum: x

S. No: 841
Ans: 256/155

Page No: 85
Sum: x

S. No: 842
Ans: 1147/825

Page No: 85
Sum: x

S. No: 843
Ans: 319/1050

Page No: 85
Sum: x

S. No: 844
Ans: 551/46

Page No: 85
Sum: x

S. No: 845
Ans: 13/6

Page No: 85
Sum: x

S. No: 846
Ans: 1995/1591

Page No: 85
Sum: x

S. No: 847
Ans: 272/247

Page No: 85
Sum: x

S. No: 848
Ans: 416/49

Page No: 85
Sum: x

S. No: 849
Ans: 1403/867

Page No: 85
Sum: x

S. No: 850
Ans: 221/2291

Page No: 86
Sum: x

S. No: 851
Ans: 295/3901

Page No: 86
Sum: x

S. No: 852
Ans: 1591/2556

Page No: 86
Sum: x

S. No: 853
Ans: 768/481

Page No: 86
Sum: x

S. No: 854
Ans: 1003/702

Page No: 86
Sum: x

S. No: 855
Ans: 602/3201

Page No: 86
Sum: x

S. No: 856
Ans: 459/155

Page No: 86
Sum: x

S. No: 857
Ans: 840/869

Page No: 86
Sum: x

S. No: 858
Ans: 1813/1320

Page No: 86
Sum: x

S. No: 859
Ans: 5035/4067

Page No: 86
Sum: x

S. No: 860
Ans: 539/1880

Page No: 87
Sum: x

S. No: 861
Ans: 7469/612

Page No: 87
Sum: x

S. No: 862
Ans: 37/140

Page No: 87
Sum: x

S. No: 863
Ans: 2212/1275

Page No: 87
Sum: x

S. No: 864
Ans: 793/952

Page No: 87
Sum: x

S. No: 865
Ans: 1209/860

Page No: 87
Sum: x

S. No: 866
Ans: 3916/1107

Page No: 87
Sum: x

S. No: 867
Ans: 7371/646

Page No: 87
Sum: x

S. No: 868
Ans: 91/152

Answers

Page No: 87
Sum: x

S. No: 869
Ans: 485/4092

Page No: 87
Sum: x

S. No: 870
Ans: 35/48

Page No: 88
Sum: x

S. No: 871
Ans: 520/441

Page No: 88
Sum: x

S. No: 872
Ans: 158/57

Page No: 88
Sum: x

S. No: 873
Ans: 2343/1940

Page No: 88
Sum: x

S. No: 874
Ans: 126/265

Page No: 88
Sum: x

S. No: 875
Ans: 57/86

Page No: 88
Sum: x

S. No: 876
Ans: 183/40

Page No: 88
Sum: x

S. No: 877
Ans: 646/1485

Page No: 88
Sum: x

S. No: 878
Ans: 2079/1037

Page No: 88
Sum: x

S. No: 879
Ans: 627/371

Page No: 88
Sum: x

S. No: 880
Ans: 154/295

Page No: 89
Sum: x

S. No: 881
Ans: 2262/893

Page No: 89
Sum: x

S. No: 882
Ans: 4539/931

Page No: 89
Sum: x

S. No: 883
Ans: 8428/7857

Page No: 89
Sum: x

S. No: 884
Ans: 124/1701

Page No: 89
Sum: x

S. No: 885
Ans: 323/88

Page No: 89
Sum: x

S. No: 886
Ans: 2278/3337

Page No: 89
Sum: x

S. No: 887
Ans: 893/4662

Page No: 89
Sum: x

S. No: 888
Ans: 118/89

Page No: 89
Sum: x

S. No: 889
Ans: 357/116

Page No: 89
Sum: x

S. No: 890
Ans: 2077/7031

Page No: 90
Sum: x

S. No: 891
Ans: 1027/1653

Page No: 90
Sum: x

S. No: 892
Ans: 665/1044

Page No: 90
Sum: x

S. No: 893
Ans: 99/686

Page No: 90
Sum: x

S. No: 894
Ans: 432/737

Page No: 90
Sum: x

S. No: 895
Ans: 456/341

Page No: 90
Sum: x

S. No: 896
Ans: 164/75

Answers

Page No: 90
Sum: x

S. No: 897
Ans: 23/20

Page No: 90
Sum: x

S. No: 898
Ans: 185/357

Page No: 90
Sum: x

S. No: 899
Ans: 1387/1265

Page No: 90
Sum: x

S. No: 900
Ans: 494/1479

Page No: 91
Sum: x

S. No: 901
Ans: 448/975

Page No: 91
Sum: x

S. No: 902
Ans: 611/186

Page No: 91
Sum: x

S. No: 903
Ans: 504/11

Page No: 91
Sum: x

S. No: 904
Ans: 1369/4416

Page No: 91
Sum: x

S. No: 905
Ans: 497/2673

Page No: 91
Sum: x

S. No: 906
Ans: 2449/768

Page No: 91
Sum: x

S. No: 907
Ans: 1127/435

Page No: 91
Sum: x

S. No: 908
Ans: 1387/507

Page No: 91
Sum: x

S. No: 909
Ans: 720/1007

Page No: 91
Sum: x

S. No: 910
Ans: 476/4005

Page No: 92
Sum: x

S. No: 911
Ans: 240/143

Page No: 92
Sum: x

S. No: 912
Ans: 511/572

Page No: 92
Sum: x

S. No: 913
Ans: 59/36

Page No: 92
Sum: x

S. No: 914
Ans: 1411/5044

Page No: 92
Sum: x

S. No: 915
Ans: 70/111

Page No: 92
Sum: x

S. No: 916
Ans: 1190/851

Page No: 92
Sum: x

S. No: 917
Ans: 3483/2068

Page No: 92
Sum: x

S. No: 918
Ans: 51/97

Page No: 92
Sum: x

S. No: 919
Ans: 1081/792

Page No: 92
Sum: x

S. No: 920
Ans: 4288/1105

Page No: 93
Sum: x

S. No: 921
Ans: 640/451

Page No: 93
Sum: x

S. No: 922
Ans: 352/153

Page No: 93
Sum: x

S. No: 923
Ans: 95/62

Page No: 93
Sum: x

S. No: 924
Ans: 187/135

Answers

Page No: 93
Sum: x

S. No: 925
Ans: 160/649

Page No: 93
Sum: x

S. No: 926
Ans: 87/70

Page No: 93
Sum: x

S. No: 927
Ans: 160/117

Page No: 93
Sum: x

S. No: 928
Ans: 451/570

Page No: 93
Sum: x

S. No: 929
Ans: 136/111

Page No: 93
Sum: x

S. No: 930
Ans: 1800/3311

Page No: 94
Sum: x

S. No: 931
Ans: 209/252

Page No: 94
Sum: x

S. No: 932
Ans: 2040/1591

Page No: 94
Sum: x

S. No: 933
Ans: 295/126

Page No: 94
Sum: x

S. No: 934
Ans: 5782/1479

Page No: 94
Sum: x

S. No: 935
Ans: 273/305

Page No: 94
Sum: x

S. No: 936
Ans: 7/184

Page No: 94
Sum: x

S. No: 937
Ans: 1292/6525

Page No: 94
Sum: x

S. No: 938
Ans: 308/159

Page No: 94
Sum: x

S. No: 939
Ans: 6076/4455

Page No: 94
Sum: x

S. No: 940
Ans: 260/291

Page No: 95
Sum: x

S. No: 941
Ans: 1079/522

Page No: 95
Sum: x

S. No: 942
Ans: 312/1015

Page No: 95
Sum: x

S. No: 943
Ans: 364/215

Page No: 95
Sum: x

S. No: 944
Ans: 188/441

Page No: 95
Sum: x

S. No: 945
Ans: 2548/589

Page No: 95
Sum: x

S. No: 946
Ans: 440/469

Page No: 95
Sum: x

S. No: 947
Ans: 81/44

Page No: 95
Sum: x

S. No: 948
Ans: 100/213

Page No: 95
Sum: x

S. No: 949
Ans: 3196/1421

Page No: 95
Sum: x

S. No: 950
Ans: 1457/885

Page No: 96
Sum: x

S. No: 951
Ans: 2720/2059

Page No: 96
Sum: x

S. No: 952
Ans: 1771/3827

Answers

Page No: 96
Sum: x

S. No: 953
Ans: 1260/781

Page No: 96
Sum: x

S. No: 954
Ans: 6975/722

Page No: 96
Sum: x

S. No: 955
Ans: 3325/2619

Page No: 96
Sum: x

S. No: 956
Ans: 264/931

Page No: 96
Sum: x

S. No: 957
Ans: 833/779

Page No: 96
Sum: x

S. No: 958
Ans: 533/1470

Page No: 96
Sum: x

S. No: 959
Ans: 4183/2440

Page No: 96
Sum: x

S. No: 960
Ans: 29/5

Page No: 97
Sum: x

S. No: 961
Ans: 387/80

Page No: 97
Sum: x

S. No: 962
Ans: 494/63

Page No: 97
Sum: x

S. No: 963
Ans: 155/1053

Page No: 97
Sum: x

S. No: 964
Ans: 81/73

Page No: 97
Sum: x

S. No: 965
Ans: 117/340

Page No: 97
Sum: x

S. No: 966
Ans: 187/399

Page No: 97
Sum: x

S. No: 967
Ans: 252/209

Page No: 97
Sum: x

S. No: 968
Ans: 6059/4628

Page No: 97
Sum: x

S. No: 969
Ans: 1679/272

Page No: 97
Sum: x

S. No: 970
Ans: 864/2795

Page No: 98
Sum: x

S. No: 971
Ans: 294/1081

Page No: 98
Sum: x

S. No: 972
Ans: 21/65

Page No: 98
Sum: x

S. No: 973
Ans: 1073/625

Page No: 98
Sum: x

S. No: 974
Ans: 2728/1275

Page No: 98
Sum: x

S. No: 975
Ans: 156/85

Page No: 98
Sum: x

S. No: 976
Ans: 1001/1340

Page No: 98
Sum: x

S. No: 977
Ans: 8648/2175

Page No: 98
Sum: x

S. No: 978
Ans: 37/154

Page No: 98
Sum: x

S. No: 979
Ans: 2665/8712

Page No: 98
Sum: x

S. No: 980
Ans: 440/2581

Answers

Page No: 99
Sum: x

S. No: 981
Ans: 4802/1935

Page No: 99
Sum: x

S. No: 982
Ans: 2584/2769

Page No: 99
Sum: x

S. No: 983
Ans: 31/7

Page No: 99
Sum: x

S. No: 984
Ans: 98/25

Page No: 99
Sum: x

S. No: 985
Ans: 475/1264

Page No: 99
Sum: x

S. No: 986
Ans: 3773/3075

Page No: 99
Sum: x

S. No: 987
Ans: 429/868

Page No: 99
Sum: x

S. No: 988
Ans: 92/119

Page No: 99
Sum: x

S. No: 989
Ans: 2537/2520

Page No: 99
Sum: x

S. No: 990
Ans: 145/258

Page No: 100
Sum: x

S. No: 991
Ans: 64/55

Page No: 100
Sum: x

S. No: 992
Ans: 473/50

Page No: 100
Sum: x

S. No: 993
Ans: 295/196

Page No: 100
Sum: x

S. No: 994
Ans: 1813/704

Page No: 100
Sum: x

S. No: 995
Ans: 2553/3395

Page No: 100
Sum: x

S. No: 996
Ans: 266/611

Page No: 100
Sum: x

S. No: 997
Ans: 3245/1863

Page No: 100
Sum: x

S. No: 998
Ans: 115/124

Page No: 100
Sum: x

S. No: 999
Ans: 2747/180

Page No: 100
Sum: x

S. No: 1000
Ans: 86/45

Page No: 101
Sum: x

S. No: 1001
Ans: 361/328

Page No: 101
Sum: x

S. No: 1002
Ans: 47/71

Page No: 101
Sum: x

S. No: 1003
Ans: 290/221

Page No: 101
Sum: x

S. No: 1004
Ans: 465/44

Page No: 101
Sum: x

S. No: 1005
Ans: 1715/3403

Page No: 101
Sum: x

S. No: 1006
Ans: 65/48

Page No: 101
Sum: x

S. No: 1007
Ans: 4

Page No: 101
Sum: x

S. No: 1008
Ans: 7280/3483

Answers

Page No: 101
Sum: x

S. No: 1009
Ans: 693/340

Page No: 101
Sum: x

S. No: 1010
Ans: 7084/3827

Page No: 102
Sum: x

S. No: 1011
Ans: 194/301

Page No: 102
Sum: x

S. No: 1012
Ans: 4753/899

Page No: 102
Sum: x

S. No: 1013
Ans: 2257/3120

Page No: 102
Sum: x

S. No: 1014
Ans: 3007/480

Page No: 102
Sum: x

S. No: 1015
Ans: 1275/1376

Page No: 102
Sum: x

S. No: 1016
Ans: 18/77

Page No: 102
Sum: x

S. No: 1017
Ans: 2376/391

Page No: 102
Sum: x

S. No: 1018
Ans: 837/143

Page No: 102
Sum: x

S. No: 1019
Ans: 209/128

Page No: 102
Sum: x

S. No: 1020
Ans: 1131/2173

Page No: 103
Sum: x

S. No: 1021
Ans: 6351/4108

Page No: 103
Sum: x

S. No: 1022
Ans: 1885/6888

Page No: 103
Sum: x

S. No: 1023
Ans: 26/85

Page No: 103
Sum: x

S. No: 1024
Ans: 3096/1225

Page No: 103
Sum: x

S. No: 1025
Ans: 64/91

Page No: 103
Sum: x

S. No: 1026
Ans: 341/4872

Page No: 103
Sum: x

S. No: 1027
Ans: 1073/368

Page No: 103
Sum: x

S. No: 1028
Ans: 767/1995

Page No: 103
Sum: x

S. No: 1029
Ans: 189/440

Page No: 103
Sum: x

S. No: 1030
Ans: 1325/4462

Page No: 104
Sum: x

S. No: 1031
Ans: 4599/3230

Page No: 104
Sum: x

S. No: 1032
Ans: 63/94

Page No: 104
Sum: x

S. No: 1033
Ans: 51/83

Page No: 104
Sum: x

S. No: 1034
Ans: 3139/2418

Page No: 104
Sum: x

S. No: 1035
Ans: 2485/726

Page No: 104
Sum: x

S. No: 1036
Ans: 2573/910

Answers

Page No: 104
Sum: x

S. No: 1037
Ans: 1075/2716

Page No: 104
Sum: x

S. No: 1038
Ans: 1462/3283

Page No: 104
Sum: x

S. No: 1039
Ans: 26/9

Page No: 104
Sum: x

S. No: 1040
Ans: 663/1504

Page No: 105
Sum: x

S. No: 1041
Ans: 46/41

Page No: 105
Sum: x

S. No: 1042
Ans: 1935/9506

Page No: 105
Sum: x

S. No: 1043
Ans: 68/21

Page No: 105
Sum: x

S. No: 1044
Ans: 513/1012

Page No: 105
Sum: x

S. No: 1045
Ans: 20/7

Page No: 105
Sum: x

S. No: 1046
Ans: 17/25

Page No: 105
Sum: x

S. No: 1047
Ans: 85/186

Page No: 105
Sum: x

S. No: 1048
Ans: 348/85

Page No: 105
Sum: x

S. No: 1049
Ans: 775/52

Page No: 105
Sum: x

S. No: 1050
Ans: 187/75

Page No: 106
Sum: ÷

S. No: 1051
Ans: 5/9

Page No: 106
Sum: ÷

S. No: 1052
Ans: 6/7

Page No: 106
Sum: ÷

S. No: 1053
Ans: 1

Page No: 106
Sum: ÷

S. No: 1054
Ans: 9/2

Page No: 106
Sum: ÷

S. No: 1055
Ans: 7/4

Page No: 106
Sum: ÷

S. No: 1056
Ans: 4/9

Page No: 106
Sum: ÷

S. No: 1057
Ans: 4/3

Page No: 106
Sum: ÷

S. No: 1058
Ans: 5/4

Page No: 106
Sum: ÷

S. No: 1059
Ans: 1

Page No: 106
Sum: ÷

S. No: 1060
Ans: 7/2

Page No: 107
Sum: ÷

S. No: 1061
Ans: 7

Page No: 107
Sum: ÷

S. No: 1062
Ans: 4

Page No: 107
Sum: ÷

S. No: 1063
Ans: 8/3

Page No: 107
Sum: ÷

S. No: 1064
Ans: 6/5

Answers

Page No: 107
Sum: ÷

S. No: 1065
Ans: 7/5

Page No: 107
Sum: ÷

S. No: 1066
Ans: 8/7

Page No: 107
Sum: ÷

S. No: 1067
Ans: 1

Page No: 107
Sum: ÷

S. No: 1068
Ans: 3/7

Page No: 107
Sum: ÷

S. No: 1069
Ans: 1

Page No: 107
Sum: ÷

S. No: 1070
Ans: 3/4

Page No: 108
Sum: ÷

S. No: 1071
Ans: 43/21

Page No: 108
Sum: ÷

S. No: 1072
Ans: 58/59

Page No: 108
Sum: ÷

S. No: 1073
Ans: 67/88

Page No: 108
Sum: ÷

S. No: 1074
Ans: 83/96

Page No: 108
Sum: ÷

S. No: 1075
Ans: 6

Page No: 108
Sum: ÷

S. No: 1076
Ans: 34/63

Page No: 108
Sum: ÷

S. No: 1077
Ans: 76/45

Page No: 108
Sum: ÷

S. No: 1078
Ans: 45/89

Page No: 108
Sum: ÷

S. No: 1079
Ans: 24/25

Page No: 108
Sum: ÷

S. No: 1080
Ans: 59/64

Page No: 109
Sum: ÷

S. No: 1081
Ans: 41/34

Page No: 109
Sum: ÷

S. No: 1082
Ans: 98/47

Page No: 109
Sum: ÷

S. No: 1083
Ans: 65/18

Page No: 109
Sum: ÷

S. No: 1084
Ans: 19/24

Page No: 109
Sum: ÷

S. No: 1085
Ans: 94/57

Page No: 109
Sum: ÷

S. No: 1086
Ans: 47/71

Page No: 109
Sum: ÷

S. No: 1087
Ans: 97/65

Page No: 109
Sum: ÷

S. No: 1088
Ans: 30/19

Page No: 109
Sum: ÷

S. No: 1089
Ans: 92/15

Page No: 109
Sum: ÷

S. No: 1090
Ans: 66/91

Page No: 110
Sum: ÷

S. No: 1091
Ans: 94/67

Page No: 110
Sum: ÷

S. No: 1092
Ans: 80/49

Answers

Page No: 110
Sum: ÷

S. No: 1093
Ans: 46/39

Page No: 110
Sum: ÷

S. No: 1094
Ans: 49/37

Page No: 110
Sum: ÷

S. No: 1095
Ans: 77/97

Page No: 110
Sum: ÷

S. No: 1096
Ans: 23/82

Page No: 110
Sum: ÷

S. No: 1097
Ans: 64/39

Page No: 110
Sum: ÷

S. No: 1098
Ans: 24/53

Page No: 110
Sum: ÷

S. No: 1099
Ans: 19/82

Page No: 110
Sum: ÷

S. No: 1100
Ans: 49/33

Page No: 111
Sum: ÷

S. No: 1101
Ans: 18/29

Page No: 111
Sum: ÷

S. No: 1102
Ans: 13/44

Page No: 111
Sum: ÷

S. No: 1103
Ans: 27/31

Page No: 111
Sum: ÷

S. No: 1104
Ans: 3/4

Page No: 111
Sum: ÷

S. No: 1105
Ans: 23/16

Page No: 111
Sum: ÷

S. No: 1106
Ans: 66/31

Page No: 111
Sum: ÷

S. No: 1107
Ans: 93/34

Page No: 111
Sum: ÷

S. No: 1108
Ans: 84/43

Page No: 111
Sum: ÷

S. No: 1109
Ans: 83/72

Page No: 111
Sum: ÷

S. No: 1110
Ans: 94/43

Page No: 112
Sum: ÷

S. No: 1111
Ans: 26/29

Page No: 112
Sum: ÷

S. No: 1112
Ans: 39/71

Page No: 112
Sum: ÷

S. No: 1113
Ans: 35/23

Page No: 112
Sum: ÷

S. No: 1114
Ans: 62/95

Page No: 112
Sum: ÷

S. No: 1115
Ans: 92/83

Page No: 112
Sum: ÷

S. No: 1116
Ans: 31/17

Page No: 112
Sum: ÷

S. No: 1117
Ans: 18/29

Page No: 112
Sum: ÷

S. No: 1118
Ans: 83/65

Page No: 112
Sum: ÷

S. No: 1119
Ans: 35/58

Page No: 112
Sum: ÷

S. No: 1120
Ans: 38/49

Answers

Page No: 113
Sum: ÷

S. No: 1121
Ans: 8/17

Page No: 113
Sum: ÷

S. No: 1122
Ans: 44/89

Page No: 113
Sum: ÷

S. No: 1123
Ans: 45/43

Page No: 113
Sum: ÷

S. No: 1124
Ans: 47/54

Page No: 113
Sum: ÷

S. No: 1125
Ans: 47/40

Page No: 113
Sum: ÷

S. No: 1126
Ans: 2/13

Page No: 113
Sum: ÷

S. No: 1127
Ans: 43/70

Page No: 113
Sum: ÷

S. No: 1128
Ans: 52/57

Page No: 113
Sum: ÷

S. No: 1129
Ans: 70/73

Page No: 113
Sum: ÷

S. No: 1130
Ans: 33/4

Page No: 114
Sum: ÷

S. No: 1131
Ans: 5/2

Page No: 114
Sum: ÷

S. No: 1132
Ans: 3/8

Page No: 114
Sum: ÷

S. No: 1133
Ans: 27/16

Page No: 114
Sum: ÷

S. No: 1134
Ans: 36

Page No: 114
Sum: ÷

S. No: 1135
Ans: 6/25

Page No: 114
Sum: ÷

S. No: 1136
Ans: 63/8

Page No: 114
Sum: ÷

S. No: 1137
Ans: 9/4

Page No: 114
Sum: ÷

S. No: 1138
Ans: 10/21

Page No: 114
Sum: ÷

S. No: 1139
Ans: 1

Page No: 114
Sum: ÷

S. No: 1140
Ans: 1/2

Page No: 115
Sum: ÷

S. No: 1141
Ans: 9/8

Page No: 115
Sum: ÷

S. No: 1142
Ans: 15/4

Page No: 115
Sum: ÷

S. No: 1143
Ans: 3/28

Page No: 115
Sum: ÷

S. No: 1144
Ans: 1/8

Page No: 115
Sum: ÷

S. No: 1145
Ans: 6/5

Page No: 115
Sum: ÷

S. No: 1146
Ans: 7/24

Page No: 115
Sum: ÷

S. No: 1147
Ans: 16

Page No: 115
Sum: ÷

S. No: 1148
Ans: 7/3

Answers

Page No: 115
Sum: ÷

S. No: 1149
Ans: 30/7

Page No: 115
Sum: ÷

S. No: 1150
Ans: 9/8

Page No: 116
Sum: ÷

S. No: 1151
Ans: 3393/6557

Page No: 116
Sum: ÷

S. No: 1152
Ans: 3355/5624

Page No: 116
Sum: ÷

S. No: 1153
Ans: 18/29

Page No: 116
Sum: ÷

S. No: 1154
Ans: 623/264

Page No: 116
Sum: ÷

S. No: 1155
Ans: 19/12

Page No: 116
Sum: ÷

S. No: 1156
Ans: 6192/4015

Page No: 116
Sum: ÷

S. No: 1157
Ans: 238/559

Page No: 116
Sum: ÷

S. No: 1158
Ans: 9/10

Page No: 116
Sum: ÷

S. No: 1159
Ans: 48/29

Page No: 116
Sum: ÷

S. No: 1160
Ans: 649/992

Page No: 117
Sum: ÷

S. No: 1161
Ans: 75/946

Page No: 117
Sum: ÷

S. No: 1162
Ans: 15/14

Page No: 117
Sum: ÷

S. No: 1163
Ans: 70/187

Page No: 117
Sum: ÷

S. No: 1164
Ans: 1739/3672

Page No: 117
Sum: ÷

S. No: 1165
Ans: 132/455

Page No: 117
Sum: ÷

S. No: 1166
Ans: 119/517

Page No: 117
Sum: ÷

S. No: 1167
Ans: 290/123

Page No: 117
Sum: ÷

S. No: 1168
Ans: 135/98

Page No: 117
Sum: ÷

S. No: 1169
Ans: 403/10

Page No: 117
Sum: ÷

S. No: 1170
Ans: 1885/1647

Page No: 118
Sum: ÷

S. No: 1171
Ans: 255/1183

Page No: 118
Sum: ÷

S. No: 1172
Ans: 2867/1500

Page No: 118
Sum: ÷

S. No: 1173
Ans: 987/712

Page No: 118
Sum: ÷

S. No: 1174
Ans: 649/475

Page No: 118
Sum: ÷

S. No: 1175
Ans: 5084/935

Page No: 118
Sum: ÷

S. No: 1176
Ans: 225/341

Answers

Page No: 118
Sum: ÷

S. No: 1177
Ans: 2451/1312

Page No: 118
Sum: ÷

S. No: 1178
Ans: 38/13

Page No: 118
Sum: ÷

S. No: 1179
Ans: 145/882

Page No: 118
Sum: ÷

S. No: 1180
Ans: 282/175

Page No: 119
Sum: ÷

S. No: 1181
Ans: 3825/3721

Page No: 119
Sum: ÷

S. No: 1182
Ans: 162/943

Page No: 119
Sum: ÷

S. No: 1183
Ans: 29/225

Page No: 119
Sum: ÷

S. No: 1184
Ans: 9/8

Page No: 119
Sum: ÷

S. No: 1185
Ans: 2451/2912

Page No: 119
Sum: ÷

S. No: 1186
Ans: 51/40

Page No: 119
Sum: ÷

S. No: 1187
Ans: 135/92

Page No: 119
Sum: ÷

S. No: 1188
Ans: 1375/637

Page No: 119
Sum: ÷

S. No: 1189
Ans: 17/50

Page No: 119
Sum: ÷

S. No: 1190
Ans: 1809/77

Page No: 120
Sum: ÷

S. No: 1191
Ans: 20/31

Page No: 120
Sum: ÷

S. No: 1192
Ans: 1023/3700

Page No: 120
Sum: ÷

S. No: 1193
Ans: 203/66

Page No: 120
Sum: ÷

S. No: 1194
Ans: 667/950

Page No: 120
Sum: ÷

S. No: 1195
Ans: 39/200

Page No: 120
Sum: ÷

S. No: 1196
Ans: 3515/903

Page No: 120
Sum: ÷

S. No: 1197
Ans: 76/221

Page No: 120
Sum: ÷

S. No: 1198
Ans: 72/13

Page No: 120
Sum: ÷

S. No: 1199
Ans: 61/69

Page No: 120
Sum: ÷

S. No: 1200
Ans: 1343/3393

Page No: 121
Sum: ÷

S. No: 1201
Ans: 2773/415

Page No: 121
Sum: ÷

S. No: 1202
Ans: 5328/3053

Page No: 121
Sum: ÷

S. No: 1203
Ans: 351/148

Page No: 121
Sum: ÷

S. No: 1204
Ans: 767/918

Answers

Page No: 121
Sum: ÷

S. No: 1205
Ans: 1358/1419

Page No: 121
Sum: ÷

S. No: 1206
Ans: 85/93

Page No: 121
Sum: ÷

S. No: 1207
Ans: 395/462

Page No: 121
Sum: ÷

S. No: 1208
Ans: 735/814

Page No: 121
Sum: ÷

S. No: 1209
Ans: 2597/7885

Page No: 121
Sum: ÷

S. No: 1210
Ans: 517/490

Page No: 122
Sum: ÷

S. No: 1211
Ans: 693/1649

Page No: 122
Sum: ÷

S. No: 1212
Ans: 185/28

Page No: 122
Sum: ÷

S. No: 1213
Ans: 84/1343

Page No: 122
Sum: ÷

S. No: 1214
Ans: 1638/1037

Page No: 122
Sum: ÷

S. No: 1215
Ans: 465/559

Page No: 122
Sum: ÷

S. No: 1216
Ans: 2403/7216

Page No: 122
Sum: ÷

S. No: 1217
Ans: 3458/1377

Page No: 122
Sum: ÷

S. No: 1218
Ans: 416/1197

Page No: 122
Sum: ÷

S. No: 1219
Ans: 880/9021

Page No: 122
Sum: ÷

S. No: 1220
Ans: 121/1680

Page No: 123
Sum: ÷

S. No: 1221
Ans: 245/234

Page No: 123
Sum: ÷

S. No: 1222
Ans: 237/38

Page No: 123
Sum: ÷

S. No: 1223
Ans: 6887/660

Page No: 123
Sum: ÷

S. No: 1224
Ans: 360/371

Page No: 123
Sum: ÷

S. No: 1225
Ans: 129/608

Page No: 123
Sum: ÷

S. No: 1226
Ans: 1734/305

Page No: 123
Sum: ÷

S. No: 1227
Ans: 561/760

Page No: 123
Sum: ÷

S. No: 1228
Ans: 4697/1836

Page No: 123
Sum: ÷

S. No: 1229
Ans: 1007/2079

Page No: 123
Sum: ÷

S. No: 1230
Ans: 385/118

Page No: 124
Sum: ÷

S. No: 1231
Ans: 2726/741

Page No: 124
Sum: ÷

S. No: 1232
Ans: 4361/969

Answers

Page No: 124
Sum: ÷

S. No: 1233
Ans: 4753/3483

Page No: 124
Sum: ÷

S. No: 1234
Ans: 108/217

Page No: 124
Sum: ÷

S. No: 1235
Ans: 187/1368

Page No: 124
Sum: ÷

S. No: 1236
Ans: 4757/6392

Page No: 124
Sum: ÷

S. No: 1237
Ans: 3478/1197

Page No: 124
Sum: ÷

S. No: 1238
Ans: 5251/512

Page No: 124
Sum: ÷

S. No: 1239
Ans: 1344/493

Page No: 124
Sum: ÷

S. No: 1240
Ans: 2449/5963

Page No: 125
Sum: ÷

S. No: 1241
Ans: 1131/1501

Page No: 125
Sum: ÷

S. No: 1242
Ans: 304/1015

Page No: 125
Sum: ÷

S. No: 1243
Ans: 14/99

Page No: 125
Sum: ÷

S. No: 1244
Ans: 67/33

Page No: 125
Sum: ÷

S. No: 1245
Ans: 209/186

Page No: 125
Sum: ÷

S. No: 1246
Ans: 41/27

Page No: 125
Sum: ÷

S. No: 1247
Ans: 144/115

Page No: 125
Sum: ÷

S. No: 1248
Ans: 255/259

Page No: 125
Sum: ÷

S. No: 1249
Ans: 1045/1679

Page No: 125
Sum: ÷

S. No: 1250
Ans: 377/1938

Page No: 126
Sum: ÷

S. No: 1251
Ans: 91/12

Page No: 126
Sum: ÷

S. No: 1252
Ans: 1457/702

Page No: 126
Sum: ÷

S. No: 1253
Ans: 77/72

Page No: 126
Sum: ÷

S. No: 1254
Ans: 12/23

Page No: 126
Sum: ÷

S. No: 1255
Ans: 781/189

Page No: 126
Sum: ÷

S. No: 1256
Ans: 237/124

Page No: 126
Sum: ÷

S. No: 1257
Ans: 345/1421

Page No: 126
Sum: ÷

S. No: 1258
Ans: 475/219

Page No: 126
Sum: ÷

S. No: 1259
Ans: 95/53

Page No: 126
Sum: ÷

S. No: 1260
Ans: 765/2492

Answers

Page No: 127
Sum: ÷

S. No: 1261
Ans: 55/156

Page No: 127
Sum: ÷

S. No: 1262
Ans: 693/949

Page No: 127
Sum: ÷

S. No: 1263
Ans: 2124/5929

Page No: 127
Sum: ÷

S. No: 1264
Ans: 8051/884

Page No: 127
Sum: ÷

S. No: 1265
Ans: 2072/5415

Page No: 127
Sum: ÷

S. No: 1266
Ans: 1955/518

Page No: 127
Sum: ÷

S. No: 1267
Ans: 2021/891

Page No: 127
Sum: ÷

S. No: 1268
Ans: 291/425

Page No: 127
Sum: ÷

S. No: 1269
Ans: 506/423

Page No: 127
Sum: ÷

S. No: 1270
Ans: 1088/4355

Page No: 128
Sum: ÷

S. No: 1271
Ans: 495/328

Page No: 128
Sum: ÷

S. No: 1272
Ans: 544/891

Page No: 128
Sum: ÷

S. No: 1273
Ans: 342/155

Page No: 128
Sum: ÷

S. No: 1274
Ans: 85/297

Page No: 128
Sum: ÷

S. No: 1275
Ans: 295/88

Page No: 128
Sum: ÷

S. No: 1276
Ans: 168/145

Page No: 128
Sum: ÷

S. No: 1277
Ans: 5/104

Page No: 128
Sum: ÷

S. No: 1278
Ans: 779/330

Page No: 128
Sum: ÷

S. No: 1279
Ans: 1323/1258

Page No: 128
Sum: ÷

S. No: 1280
Ans: 86/77

Page No: 129
Sum: ÷

S. No: 1281
Ans: 133/99

Page No: 129
Sum: ÷

S. No: 1282
Ans: 555/1462

Page No: 129
Sum: ÷

S. No: 1283
Ans: 118/35

Page No: 129
Sum: ÷

S. No: 1284
Ans: 1711/4998

Page No: 129
Sum: ÷

S. No: 1285
Ans: 455/183

Page No: 129
Sum: ÷

S. No: 1286
Ans: 56/23

Page No: 129
Sum: ÷

S. No: 1287
Ans: 493/1900

Page No: 129
Sum: ÷

S. No: 1288
Ans: 583/525

Answers

Page No: 129
Sum: ÷

S. No: 1289
Ans: 155/539

Page No: 129
Sum: ÷

S. No: 1290
Ans: 1261/240

Page No: 130
Sum: ÷

S. No: 1291
Ans: 234/2407

Page No: 130
Sum: ÷

S. No: 1292
Ans: 840/377

Page No: 130
Sum: ÷

S. No: 1293
Ans: 560/559

Page No: 130
Sum: ÷

S. No: 1294
Ans: 196/423

Page No: 130
Sum: ÷

S. No: 1295
Ans: 931/1612

Page No: 130
Sum: ÷

S. No: 1296
Ans: 70/737

Page No: 130
Sum: ÷

S. No: 1297
Ans: 16/11

Page No: 130
Sum: ÷

S. No: 1298
Ans: 48/71

Page No: 130
Sum: ÷

S. No: 1299
Ans: 833/1363

Page No: 130
Sum: ÷

S. No: 1300
Ans: 4185/2773

Page No: 131
Sum: ÷

S. No: 1301
Ans: 6035/3712

Page No: 131
Sum: ÷

S. No: 1302
Ans: 2047/3311

Page No: 131
Sum: ÷

S. No: 1303
Ans: 495/497

Page No: 131
Sum: ÷

S. No: 1304
Ans: 25/62

Page No: 131
Sum: ÷

S. No: 1305
Ans: 1843/189

Page No: 131
Sum: ÷

S. No: 1306
Ans: 456/539

Page No: 131
Sum: ÷

S. No: 1307
Ans: 697/931

Page No: 131
Sum: ÷

S. No: 1308
Ans: 2009/390

Page No: 131
Sum: ÷

S. No: 1309
Ans: 2867/3560

Page No: 131
Sum: ÷

S. No: 1310
Ans: 29/45

Page No: 132
Sum: ÷

S. No: 1311
Ans: 45/172

Page No: 132
Sum: ÷

S. No: 1312
Ans: 1053/266

Page No: 132
Sum: ÷

S. No: 1313
Ans: 405/403

Page No: 132
Sum: ÷

S. No: 1314
Ans: 5913/1156

Page No: 132
Sum: ÷

S. No: 1315
Ans: 65/612

Page No: 132
Sum: ÷

S. No: 1316
Ans: 8075/3696

Answers

Page No: 132
Sum: ÷

S. No: 1317
Ans: 3773/1539

Page No: 132
Sum: ÷

S. No: 1318
Ans: 6497/4316

Page No: 132
Sum: ÷

S. No: 1319
Ans: 828/1241

Page No: 132
Sum: ÷

S. No: 1320
Ans: 195/344

Page No: 133
Sum: ÷

S. No: 1321
Ans: 2303/1242

Page No: 133
Sum: ÷

S. No: 1322
Ans: 135/91

Page No: 133
Sum: ÷

S. No: 1323
Ans: 148/261

Page No: 133
Sum: ÷

S. No: 1324
Ans: 775/1122

Page No: 133
Sum: ÷

S. No: 1325
Ans: 65/51

Page No: 133
Sum: ÷

S. No: 1326
Ans: 5159/260

Page No: 133
Sum: ÷

S. No: 1327
Ans: 4002/1175

Page No: 133
Sum: ÷

S. No: 1328
Ans: 2849/1250

Page No: 133
Sum: ÷

S. No: 1329
Ans: 585/328

Page No: 133
Sum: ÷

S. No: 1330
Ans: 2552/445

Page No: 134
Sum: ÷

S. No: 1331
Ans: 43/90

Page No: 134
Sum: ÷

S. No: 1332
Ans: 1349/1326

Page No: 134
Sum: ÷

S. No: 1333
Ans: 279/175

Page No: 134
Sum: ÷

S. No: 1334
Ans: 225/98

Page No: 134
Sum: ÷

S. No: 1335
Ans: 3600/1501

Page No: 134
Sum: ÷

S. No: 1336
Ans: 1148/825

Page No: 134
Sum: ÷

S. No: 1337
Ans: 1209/308

Page No: 134
Sum: ÷

S. No: 1338
Ans: 272/161

Page No: 134
Sum: ÷

S. No: 1339
Ans: 413/430

Page No: 134
Sum: ÷

S. No: 1340
Ans: 750/1247

Page No: 135
Sum: ÷

S. No: 1341
Ans: 44/45

Page No: 135
Sum: ÷

S. No: 1342
Ans: 539/86

Page No: 135
Sum: ÷

S. No: 1343
Ans: 236/245

Page No: 135
Sum: ÷

S. No: 1344
Ans: 592/539

Answers

Page No: 135
Sum: ÷

S. No: 1345
Ans: 2415/3589

Page No: 135
Sum: ÷

S. No: 1346
Ans: 893/182

Page No: 135
Sum: ÷

S. No: 1347
Ans: 495/1357

Page No: 135
Sum: ÷

S. No: 1348
Ans: 713/20

Page No: 135
Sum: ÷

S. No: 1349
Ans: 603/205

Page No: 135
Sum: ÷

S. No: 1350
Ans: 160/387

Page No: 136
Sum: ÷

S. No: 1351
Ans: 72/41

Page No: 136
Sum: ÷

S. No: 1352
Ans: 144/3337

Page No: 136
Sum: ÷

S. No: 1353
Ans: 754/85

Page No: 136
Sum: ÷

S. No: 1354
Ans: 1023/7220

Page No: 136
Sum: ÷

S. No: 1355
Ans: 287/415

Page No: 136
Sum: ÷

S. No: 1356
Ans: 320/351

Page No: 136
Sum: ÷

S. No: 1357
Ans: 121/49

Page No: 136
Sum: ÷

S. No: 1358
Ans: 6480/3913

Page No: 136
Sum: ÷

S. No: 1359
Ans: 612/385

Page No: 136
Sum: ÷

S. No: 1360
Ans: 6853/3956

Page No: 137
Sum: ÷

S. No: 1361
Ans: 344/679

Page No: 137
Sum: ÷

S. No: 1362
Ans: 1421/3007

Page No: 137
Sum: ÷

S. No: 1363
Ans: 481/915

Page No: 137
Sum: ÷

S. No: 1364
Ans: 776/4185

Page No: 137
Sum: ÷

S. No: 1365
Ans: 600/731

Page No: 137
Sum: ÷

S. No: 1366
Ans: 22/63

Page No: 137
Sum: ÷

S. No: 1367
Ans: 736/561

Page No: 137
Sum: ÷

S. No: 1368
Ans: 341/39

Page No: 137
Sum: ÷

S. No: 1369
Ans: 152/99

Page No: 137
Sum: ÷

S. No: 1370
Ans: 4611/2132

Page No: 138
Sum: ÷

S. No: 1371
Ans: 3796/6873

Page No: 138
Sum: ÷

S. No: 1372
Ans: 1218/2665

Answers

Page No: 138
Sum: ÷

S. No: 1373
Ans: 85/26

Page No: 138
Sum: ÷

S. No: 1374
Ans: 2107/1800

Page No: 138
Sum: ÷

S. No: 1375
Ans: 361/1456

Page No: 138
Sum: ÷

S. No: 1376
Ans: 616/2697

Page No: 138
Sum: ÷

S. No: 1377
Ans: 37/667

Page No: 138
Sum: ÷

S. No: 1378
Ans: 455/3363

Page No: 138
Sum: ÷

S. No: 1379
Ans: 280/33

Page No: 138
Sum: ÷

S. No: 1380
Ans: 2438/2425

Page No: 139
Sum: ÷

S. No: 1381
Ans: 2142/6935

Page No: 139
Sum: ÷

S. No: 1382
Ans: 423/224

Page No: 139
Sum: ÷

S. No: 1383
Ans: 1411/1587

Page No: 139
Sum: ÷

S. No: 1384
Ans: 3999/1898

Page No: 139
Sum: ÷

S. No: 1385
Ans: 105/142

Page No: 139
Sum: ÷

S. No: 1386
Ans: 2821/830

Page No: 139
Sum: ÷

S. No: 1387
Ans: 2425/1204

Page No: 139
Sum: ÷

S. No: 1388
Ans: 1139/4214

Page No: 139
Sum: ÷

S. No: 1389
Ans: 13/72

Page No: 139
Sum: ÷

S. No: 1390
Ans: 1833/544

Page No: 140
Sum: ÷

S. No: 1391
Ans: 46/41

Page No: 140
Sum: ÷

S. No: 1392
Ans: 4214/4365

Page No: 140
Sum: ÷

S. No: 1393
Ans: 68/21

Page No: 140
Sum: ÷

S. No: 1394
Ans: 5589/3344

Page No: 140
Sum: ÷

S. No: 1395
Ans: 7/80

Page No: 140
Sum: ÷

S. No: 1396
Ans: 25/68

Page No: 140
Sum: ÷

S. No: 1397
Ans: 510/31

Page No: 140
Sum: ÷

S. No: 1398
Ans: 29/255

Page No: 140
Sum: ÷

S. No: 1399
Ans: 1116/325

Page No: 140
Sum: ÷

S. No: 1400
Ans: 275/51

Made in the USA
Monee, IL
02 November 2026

67374654R00109